Nuclear Power

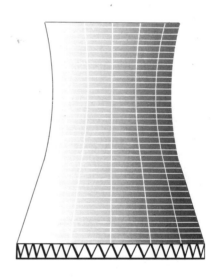

Jonathan F. Gosse

AMERICAN TECHNICAL PUBLISHERS, INC.
HOMEWOOD, ILLINOIS 60430

1 2 3 4 5 6 7 8 9 -90- 9 8 7 6 5 4 3 2 1

Printed in the United States of America

Library of Congress Cataloging-in-Publication Data

Gosse, Jonathan., 1952–
  Nuclear power / Jonathan F. Gosse.
    p. cm.
  ISBN 0-8269-3406-4 : $16.96
  1. Nuclear power plants.   2. Nuclear power plants—Safety
measures.   3. Nuclear fuels.   4. Nuclear energy.   I. Title.
TK1078.G68   1990
621.48'3—dc20   90-365
                                                          CIP

Contents

Acknowledgments

The author and publisher are grateful to the following companies, organizations, and individuals for providing technical information and assistance.

Brookhaven National Laboratory
Dosimeter Corporation
Duke Power Co.
Eberline Instruments
Food and Drug Administration
GE Medical Systems
General Electric Company
Ludlum Measurements, Inc.
Maritime Administration
The Martin Co.
Mount Wilson and Palomar Observatories
NASA
National Institutes of Health
Oak Ridge National Laboratory
Parke, Davis and Company
Public Service Company of Colorado, Fort St. Vrain
Radiation Dynamics, Inc.
Tech/Ops Landauer, Inc.
Tennessee Valley Authority
U.S. Council for Energy Awareness
U.S. Navy
Victoreen, Inc.
Washington Public Power Supply System
Westinghouse Hanford Company
Westinghouse Savannah River Company
Zurn Construction, Inc./Balcke Dürr, AG

Winston H. Heneveld, formerly Principal Engineer
Rockwell Hanford Operations
Rockwell International

Bill E. Wilcoxson, Business Manager-Financial Secretary
Local #112
International Brotherhood of Electrical Workers

Introduction

Nuclear power has an increasing role in many aspects of our daily lives. A major role of nuclear power is in the generation of electricity for industrial, commercial, and residential use. Additionally, nuclear power is used in medical, scientific, agricultural, military, and space exploration applications. Today, well over 100 nuclear power plants are operating in the United States.

NUCLEAR POWER presents basic information essential for tradesworkers employed in the construction, maintenance, operation, and decommissioning of nuclear power plant facilities. Nuclear history, theory, measurement, reactor principles, security, dosimetry, safety, and radioactive waste storage are presented and illustrated.

Appendices in **NUCLEAR POWER** provide reference tables, nuclear power acronyms, and additional information sources. The illustrated glossary defines key terms related to nuclear power.

This book contains procedures practiced in the nuclear power industry. For maximum safety, always refer to specific plant procedures and applicable federal, state, and local regulations.

Chapter 1

Nuclear Energy Principles

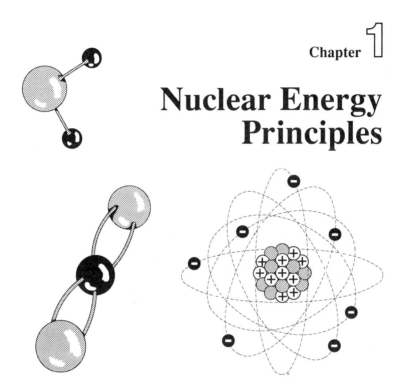

Nuclear energy is the energy released from an atomic reaction. Research and experimentation in nuclear energy has occurred over many years as scientists have searched for new sources of energy to replace rapidly decreasing fossil fuels such as coal and oil.

The atom is the smallest building block of matter that cannot be divided into smaller units without changing its basic character. Atoms are composed of neutrons, protons, and electrons in varying numbers to make the 89 naturally occurring and 20 artificially produced atoms that comprise the known chemical elements. Nuclear fission is the energy released when an atom is split. This energy, first used in the atom bomb near the end of World War II, is today being harnessed to produce electricity for residential, commercial, and industrial applications. It continues to be used for military purposes and is also used in medicine and the space program.

NUCLEAR ENERGY THEORY

Nuclear energy is the energy released from an atomic reaction. *Nuclear fission* is the energy released when an atom is split into two atoms. *Nuclear fusion* is energy released when two atoms are forced together to make a third atom. See Figure 1-1. Nuclear fission releases more energy that can be used for heat than any other energy source currently available. Nuclear fission can be used to produce heat for the same industrial purposes as fossil fuels such as coal and petroleum products. Nuclear fusion is believed to be the source of energy for the sun. Recently there has been speculation regarding the ability to create nuclear fusion. However, experimentation with nuclear fusion has not produced conclusive results that can be used by industry.

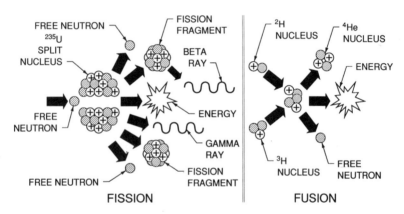

Figure 1-1. Nuclear energy is the energy released from an atomic reaction.

Radioactivity

Materials that release nuclear energy in the form of penetrating rays are radioactive, or contain radioactivity. Some materials contain certain atoms known as radioisotopes, which also release radioactivity. Radioisotopes are created from the splitting of atoms, or from atoms that have absorbed large amounts of energy during fission. This energy is then released in the form of rays.

Exposure from radioactive material depends on the source, intensity, length of exposure, and distance from the source. See Figure 1-2.

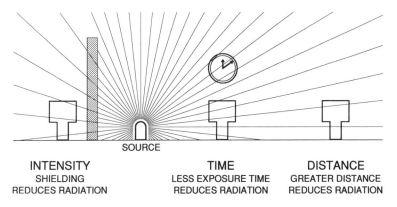

INTENSITY
SHIELDING
REDUCES RADIATION

TIME
LESS EXPOSURE TIME
REDUCES RADIATION

DISTANCE
GREATER DISTANCE
REDUCES RADIATION

Figure 1-2. Materials that release nuclear energy are radioactive.

MATTER

All matter is composed of chemical elements. *Chemical elements* are substances that cannot be broken down into simpler substances. There are 109 chemical elements (20 are artificial, 89 are natural) that cannot be broken down into simpler substances. For example, chemical elements such as iron and lead contain atoms of only one type. See Appendix. *Chemical compounds* are substances containing atoms of two or more different chemical elements. A chemical compound always has one definite composition based on the atoms contained in the chemical compound. For example, water always has two hydrogen atoms and one oxygen atom, never more or less. Chemical mixtures are different from chemical compounds. *Chemical mixtures* are mixtures of chemical elements and/or chemical compounds. For example, air is a chemical mixture of nitrogen, oxygen, carbon dioxide, and other gases. The chemical mixture of air can change based on the amount of chemical elements present.

Chemical elements are classified according to their weight (mass). Of the natural chemical elements, hydrogen is the lightest. Uranium is the heaviest natural chemical element. The mass of a chemical element is determined by the atoms that make up the chemical element. See Figure 1-3.

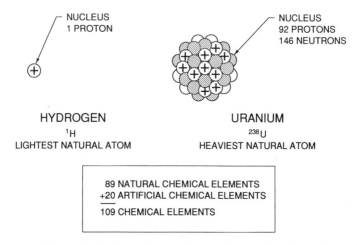

Figure 1-3. Chemical elements are classified according to their weight (mass).

ATOMS

Atoms are the smallest building blocks of matter that cannot be divided into smaller units without changing their basic character. The *nucleus*, which contains protons and neutrons, is the heavy, dense center of the atom and has a positive electrical charge. The nucleus is surrounded by one or more electrons. *Electrons* are negative-charged particles of matter that orbit around the nucleus. See Figure 1-4. It is difficult to imagine how small an atom is. In addition to being very small, atoms consist primarily of empty space. For example, if the nucleus of a hydrogen atom could be enlarged to the size of a baseball, the closest electron to the nucleus would be eight blocks away. Another example is if a marble representing a nucleus were placed on the 50-yard line of The Kingdome in Seattle, Washington, electrons would be orbiting at roof level. If all empty space were removed from an atom, pure, condensed matter would be produced. Pure, condensed matter the size of a drop of water would weigh two million tons.

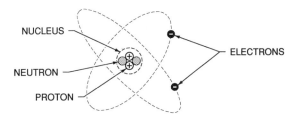

Figure 1-4. Atoms are the smallest building blocks of matter that cannot be divided into smaller units without changing their basic character.

Nucleus

The nucleus contains a cluster of two kinds of particles, protons and neutrons, which are similar in size and weight. A *proton* is a particle with a positive electrical charge. A *neutron* is a particle with no electrical charge. The difference between the two particles is easy to remember because the word neutron sounds similar to electrically neutral. See Figure 1-5.

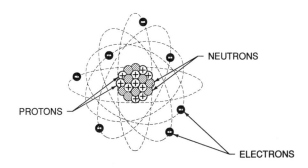

Figure 1-5. Protons contain a positive electrical charge. Neutrons contain no electrical charge. Electrons contain a negative electrical charge.

Electrons

Negatively charged electrons orbit around the nucleus of the atom. Electrons are similar to satellites orbiting around the Earth. They are much lighter than neutrons or protons. The weight (mass) of a neutron or proton is about 1.7×10^{-24} grams. The mass of an electron is 9.1×10^{-28} grams or approximately $\frac{1}{1840}$ the mass of a proton or neutron. The entire atom is 10,000 times larger than the nucleus. There are six sextillion (6×10^{21}) atoms in a drop of water.

The negative electrical charge of an electron is equally as strong as the positive electrical charge of a proton. When the atom is in its normal state, the number of negatively charged electrons orbiting around the nucleus is the same as the number of positively charged protons in the nucleus. See Figure 1-6. Because one electron neutralizes the charge of one proton, the atom as a whole has neither a positive nor negative electrical charge. Hydrogen, the lightest element, has one proton in the nucleus and one electron. Hydrogen has no neutron in the nucleus. All other chemical elements have a neutron in the nucleus.

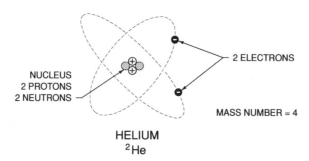

Figure 1-6. The nucleus of an atom contains protons and neutrons.

The mass of a chemical element is determined by the number of protons and neutrons in the nucleus. The heavier the chemical element, the more protons and neutrons will be present in the nucleus. Protons have a positive electrical charge and neutrons have no electrical charge. *Nuclear forces* are the powerful forces that hold the nucleus together. *Binding energy* is the energy required to overcome nuclear forces to release protons and neutrons. See Figure 1-7.

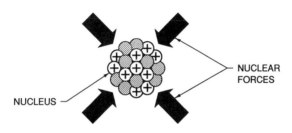

Figure 1-7. The nucleus of an atom is held together by nuclear forces.

ISOTOPES

Chemical elements can have more than one form based on the number of neutrons in the nucleus. The *isotope* is the form of the chemical element. Isotopes are classified by their mass number. The *mass number* is the total number of protons and neutrons in the nucleus. All isotopes of the same chemical element have the same number of protons in the nucleus. For example, hydrogen always has just one proton in its nucleus. Uranium always has 92 protons in its nucleus. Isotopes of a chemical element are designated by numbers after the element name that describe the total number of protons and neutrons present in the nucleus.

The mass number specifies the total number of protons and neutrons in an atom and is used to identify the specific isotope. See Figure 1-8. For example, uranium 238 (^{238}U) has a mass number of 238. The mass number of 238 is the sum of 92 protons and 146 neutrons. The most commonly used uranium isotope in nuclear power plants is uranium 235 (^{235}U). Uranium 235 consists of 92 protons and 143 neutrons. Although isotopes of a chemical element behave the same chemically, they can vary in other properties. For example, carbon 14 (^{14}C) is radioactive,

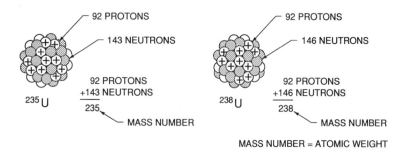

MASS NUMBER = ATOMIC WEIGHT

Figure 1-8. The mass number of an atom is the sum of the number of protons and neutrons in the nucleus.

making it a radioisotope. Isotopes of a chemical element will have different numbers of neutrons. However, no two different chemical elements will ever have the same number of protons in their atoms.

Molecules

Molecules are two or more atoms joined together by forces. The atoms in a molecule can be all the same or different. For example, one molecule of water (H_2O) contains two atoms of hydrogen and one atom of oxygen. One molecule of carbon dioxide (CO_2) contains two atoms of oxygen and one atom of carbon. See Figure 1-9.

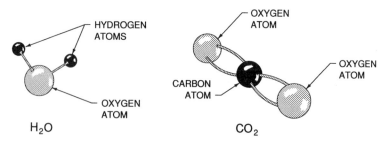

Figure 1-9. Molecules are two or more atoms joined together.

SCIENTIFIC NOTATION

Measurements of the parts of atoms are similar to measurements used in astronomy. The size of the measurements often makes it difficult to read and write the numbers accurately. *Scientific notation* is the process of using powers of 10 to simplify numbers such as .00000000134 and 32,000,000,000,000. For example, instead of writing 100, 100 can be expressed as $1 \times 10 \times 10$. This is also equivalent to 10^2. The number is now read one times ten to the second power. The number 2 is the exponent. An *exponent* indicates a power of 10 in scientific notation.

Using scientific notation, 400,000,000,000 can be written as 4×10^{11}. The number is now read four times ten to the eleventh power. The exponent is 11. The exponent also indicates how many places to move the decimal point in 400,000,000,000 to get 4. In this case, the decimal point is moved 11 places to the right. See Figure 1-10.

POWERS OF 10	
1×10^4 = 10,000	= $10 \times 10 \times 10 \times 10$ (Read ten to the fourth power)
1×10^3 = 1000	= $10 \times 10 \times 10$ (Read ten to the third power or 10 cubed)
1×10^2 = 100	= 10×10 (Read ten to the second power or 10 squared)
1×10^1 = 10	= 10 (Read ten to the first power)
1×10^0 = 1	= 1 (Read ten to the zero power)
1×10^{-1} = .1	= 1/10 (Read ten to the minus first power)
1×10^{-2} = .01	= $1/(10 \times 10)$ or 1/100 (Read ten to the minus second power)
1×10^{-3} = .001	= $1/(10 \times 10 \times 10)$ or 1/1000 (Read ten to the minus third power)
1×10^{-4} = .0001	= $1/(10 \times 10 \times 10 \times 10)$ or 1/10,000 (Read ten to the minus fourth power)

Figure 1-10. Using the powers of 10 simplifies math for very large or small numbers.

Decimal Point

In scientific notation, 10^2 means that the decimal point is moved two places to the right in order to obtain the original number. In a similar fashion, 10^{-2} means the decimal point is moved two places to the left. When a number is written with no decimal point, it means that the decimal point comes at that end. For example, 1×10^2 equals 100, and 1×10^{-2} equals .01. In all cases, an exponent with the minus sign means a smaller number.

Scientific notation is also used with more complex numbers such as 3,820,000,000 or .01304. Using scientific notation, these numbers are expressed as a number between 1 and 10 (including 1) multiplied by a power of 10. The best method to use when expressing complex numbers using scientific notation is to move the decimal point until a number between 1 and 10 (including 1) is obtained. For example, with the number 3,820,000,000, each time the deci- mal point is moved one place to the left it has the same effect as dividing the number by 10 (3,820,000,000 = 382,000,000.0 $\times$ 10 = 38,200,000.00 $\times$ 10^2 = 3,820,000.000 $\times$ 10^3). This process continues until 3.82×10^9 is obtained. To obtain 3.82 out of the original number, the decimal is moved to the left nine places. The final answer is 3.82×10^9. To express .01304 in scientific notation, a number between 1 and 10 (including 1) must also be obtained. This is accomplished by moving the decimal point two places to the right or 1.304 $\times$ 10 $\times$ 10^{-2}. The reason a number between 1 and 10 (including 1) is used in scientific notation is that multiplication and division are easier to perform if all the numbers are written in the same form.

Addition and Subtraction

To add or subtract numbers using scientific notation, the numbers must have the same exponents. For example, 5,230,000 + 326,000,000 = 5.23×10^6 + 326×10^6 = 331.23×10^6 = 331,230,000. This same method is used when subtracting. For example, 97,000,000,000 – 13,000,000 = 9.7×10^{10} – 1.3×10^7.

The exponents must be the same; therefore, $9700 \times 10^7 - 1.3 \times 10^7 = 9698.7 \times 10^7$.

Multiplication and Division

To multiply numbers using scientific notation, the exponents of the two numbers are added together. For example, $100 \times 100 = 10,000$. Using scientific notation, this problem is written as $1 \times 10^2 \times 1 \times 10^2 = 1 \times 10^4$. The exponents total 4. The answer is 1×10^4.

To multiply $28.3 \times 30,000,000,000$, scientific notation is used to express the problem as 2.83×10^1 and 3×10^{10} ($2.83 \times 10^1 \times 3 \times 10^{10} = 8.49 \times 10^{11}$). This answer is obtained by multiplying 2.83×3 and adding the exponents 1 and 10. Three or more numbers can be multiplied using the same method by multiplying all the numbers, and adding all the exponents. For example, $28.3 \times 30,000,000,000 \times 30,000,000$ is expressed as $2.83 \times 10^1 \times 3 \times 10^{10} \times 3 \times 10^7 = (2.83 \times 3 \times 3) \times (10^1 \times 10^{10} \times 10^7) = 25.47 \times 10^{18}$.

It is difficult to multiply these numbers quickly without the use of a calculator. Also, it is easy to confuse the correct number of zeros. This method also can be used for numbers less than zero. For example, $.00013 \times .00432$ is expressed in scientific notation as $1.3 \times 10^{-4} \times 4.32 \times 10^{-3} = 5.616 \times 10^{-7}$. Multiplying $.00013 \times 4320$ converts to $1.3 \times 10^{-4} \times 4.32 \times 10^3 = 5.616 \times 10^{-1}$. When -4 and 3 are added, the answer is -1.

Division is accomplished using the same method as multiplication, except the numbers are divided, and the exponents are subtracted instead of added. For example, $45,000,000$ divided by $15,000$ is expressed in scientific notation as

$$\frac{4.5 \times 10^7}{1.5 \times 10^4} = \frac{4.5}{1.5} \times 10^3 = 3 \times 10^3$$

When $45,000,000$ is divided by $.0015$, it is expressed in scientific notation as

$$\frac{4.5 \times 10^7}{1.5 \times 10^{-3}} = \frac{4.5}{1.5} \times 10^{10} = 3 \times 10^{10}$$

When –3 is subtracted from 7, the answer is 10.

Estimation

Estimation using scientific notation allows a person to quickly approximate answers to a problem with a great degree of accuracy. For example, $28,793 \times 403,107$ appears to be a complex problem. However, by using approximate numbers and scientific notation, the answer can be quickly estimated. The number 28,973 is rounded up to 30,000 and is expressed in scientific notation as 3×10^4. The number 403,107 is rounded down to 400,000 and is expressed in scientific notation as 4×10^5. Using multiplication, $3 \times 10^4 \times 4 \times 10^5 = 12 \times 10^9$, or 1.2×10^{10}. The estimated answer is 1.2×10^{10}, or 12,000,000,000. The actual answer is 11,606,659,851.

NUCLEAR ENERGY HISTORY

The commercial benefits of nuclear energy have been realized just within the last 50 years. However, the experimentation in nuclear energy has occurred over many years. Scientists have continually searched for new sources of energy. This search has led scientists to investigate the most basic building blocks of all matter.

Research into the atomic particles had its beginnings in the experiments of the English physicist Michael Faraday in 1834. His research was concerned with the passage of electricity through solutions, and showed that each molecule in solution carries a constant amount of electrical charge. However, measurements on free particles separated from their atoms were not made until 1897 by J. J. Thomson. His research revealed that free particles removed from atoms had a negative electrical charge. The properties of the particles were the same no matter what kind of atom they were removed from. The particles identified by Thomson were what are now called electrons.

Because electrons had a negative electrical charge, it was assumed that there must be an equal and opposite positive electrical charge to balance them. In 1911, Ernest Rutherford proposed that particles with a positive electrical charge (protons) were contained in the center or nucleus of the atom. In addition, he also proposed that practically all the weight of the atom was contained in the nucleus. In 1932, James Chadwick discovered that in addition to protons, heavy neutral particles (neutrons) were also contained in the nucleus of the atom.

Another major discovery occurred in 1895 when Wilhelm Roentgen discovered the penetrating radiation he called "X rays." An *X ray* is an electromagnetic radiation with a very short wavelength. Roentgen was experimenting with the passage of electricity through a glass tube from which the air had been evacuated. When he turned off the lights he noticed a greenish glow from a paper screen that had been coated with barium cyanide crystals. This glow was caused by rays coming from the glass tube and hitting the paper screen. The rays came from the part of the tube being struck by electrons passing through the tube. See Figure 1-11.

Parke, Davis and Company

Figure 1-11. X rays were first produced by passing electricity through a vacuum tube.

The property of X rays that makes them useful is that they pass through some materials more easily than through others. Roentgen learned of this ability by accident. Experimenting to see what effect the rays would have on a photographic film, he wrapped a film in paper and laid a key on top to hold it down. After the film was exposed to the rays and developed, he found that he had a perfect outline of the key. The rays had passed through the paper but they would not pass through the dense metal of the key.

A few years later Henri Becquerel discovered that uranium emitted rays similar to X rays. Marie and Pierre Curie discovered radium, which was about a million times more active in emitting rays than uranium. Both of these discoveries indicated that some sort of reaction was taking place in the atoms of certain metals. This reaction produced energy and changed the atoms into different metals. Rutherford was the first to theorize that some of the radiation of uranium and radium consisted of particles emitted from the nucleus. This caused the nucleus to change into an entirely different element.

Nuclear Reactions

After it was discovered that the natural release of particles from one nucleus could change the nucleus of other atoms to different elements, the next step was to try to change a nucleus from one type to another artificially. This was to be accomplished by bombarding a nucleus with high-speed nuclear particles that had enough energy to penetrate the nucleus of the target atom. However, the only nuclear particles available shortly after World War I were those that were emitted from the naturally occurring unstable nuclei, such as radium. These nuclei emitted alpha particles. *Alpha particles* are positive-charged nuclear particles identical to a helium atom, which has two protons and two neutrons. An alpha particle is ejected at high speed in certain radioactive transformations.

In 1919, Rutherford bombarded nitrogen with alpha particles and caused the nitrogen to be transmuted (changed) into oxygen.

Along with the oxygen a proton was generated. This proton had more energy than the alpha particle that produced the reaction. This fact confirmed the theory that extra energy had been released to the proton. The extra energy came from the nucleus of nitrogen after the alpha particle had combined with it. This produced a nuclear reaction with the release of a large amount of energy artificially for the first time. In addition to fission and fusion, other nuclear reactions can result in the release of energy from the nucleus.

First Atom Bomb

In 1938, two scientists in Nazi Germany, Otto Hahn and Fritz Strassman, succeeded in splitting uranium atoms. Each atom of uranium was split into a barium atom and a krypton atom. After splitting from uranium, these two atoms together weighed a little less than the uranium atom. The remaining part of the uranium atom was converted into energy. The calculations that proved this were made by Lise Meitner, a refugee from Nazi Germany who had fled to Denmark.

The results of this experiment were carried to physicist Albert Einstein. Einstein, born in Germany, made contributions to science that led to many discoveries in atomic energy. The equation $E = mc^2$ (energy = mass × speed of light squared) was instrumental in the development of nuclear energy. Einstein studied in Switzerland and was deeply involved with research regarding molecular reactions in Switzerland and Germany. His professional achievements led to a position at the Institute for Advanced Study at Princeton University in New Jersey. Einstein accepted the position because of a need to work unrestrained or not harassed by the Nazi Government in Germany. A group of American scientists realized it might be possible to build a bomb that would release the tremendous energies available from nuclear fission. The scientists asked Einstein to write to President Franklin D. Roosevelt and explain the terrible consequences that would follow if America's enemies were first to build an atom bomb. Considering the brilliant work that had been done by

German scientists in nuclear physics, it seemed likely that Germany would be first to produce such a bomb.

Manhattan Project

The Manhattan Project was formed by the U.S. government in 1942 to develop the first atom bomb. This project resulted because of Einstein's warning to President Roosevelt about the possibility of German scientists developing the atom bomb. The Manhattan Project was a gamble as it had not been proven that an atom bomb was possible. Nazi scientists had decided an atom bomb was not possible and dropped their project. If they had not dropped the project, it is possible that Germany could have developed the atom bomb before the United States.

One of the key experiments designed to show whether an atom bomb could be built was conducted by Enrico Fermi under the football stands at the University of Chicago's Stagg Field on December 2, 1942. Although it was generally accepted by physicists that splitting the atom would release tremendous energy, nobody knew whether such a reaction could be made to continue long enough to be of practical use or could be controlled.

To prove the possibility of controlled nuclear fission, Fermi and his staff designed a pile of graphite blocks with uranium rods running through them. See Figure 1-12. According to their calculations, each time a uranium atom in the pile split, it would release neutrons that would strike other uranium atoms and cause them to split. These atoms would then send out more neutrons that would split other atoms. The process would repeat itself, resulting in a chain reaction. A *chain reaction* is the process in which released neutrons from an atom strike and split other atoms, which repeats the procedure. Fission is maintained in a chain reaction. In Fermi's experiment, the chain reaction would continue as long as the fissionable uranium lasted. The scientists also believed that if the chain reaction worked and could be controlled, the atom bomb and other uses of nuclear energy could be possible.

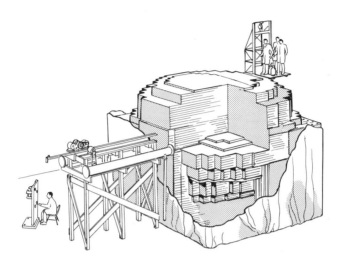

Figure 1-12. The first controlled nuclear chain reaction occurred in December, 1942.

The physicists designed the pile of graphite blocks according to careful mathematical calculations so that it would support a nuclear fission chain reaction that could be started and stopped at will. Because the graphite blocks resembled nothing more than a pile of bricks, it was called a "pile." Since that time, the term "pile" has been used to refer to a nuclear reactor.

While early nuclear research in the United States involved the development of an atom bomb, many of the scientists working on the Manhattan Project believed nuclear power could be harnessed for peaceful purposes. For example, Fermi wrote, "We all hoped that with the end of the war, emphasis would be shifted directly from the weapon to the peaceful aspects of atomic energy."

After World War II, the U.S. government promoted the development of nuclear power for civilian use. The first light water reactor constructed to produce electricity was commissioned in 1957. Projects that followed were joint efforts of government and the private sector.

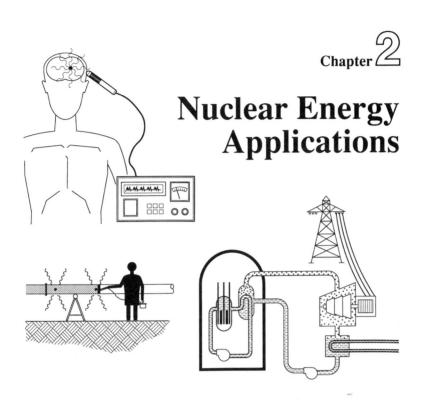

Chapter 2

Nuclear Energy Applications

Power plants that produce nuclear power are stationary or portable. Stationary power plants are generally located near a source of water. Portable power plants are used in remote locations on Earth and in space. The energy produced in nuclear power plants has many applications. Today, it is used to produce over 16% of the electricity needed for residential, commercial, and industrial applications.

Naval vessels use nuclear energy to provide power for ships and submarines. The space program uses nuclear energy in a variety of ways. Nuclear weapons are produced for military uses. The medical field uses radioisotopes (atoms that emit radiation) and radiotracers (radioisotopes that trace specific chemical elements) in medical procedures and research. Scientists use radioisotopes to date materials. Industry uses nuclear energy in manufacturing and quality control. The demand for nuclear energy is increasing because it can be used in so many ways.

POWER PLANTS

Power plants generate electricity. They require basic energy sources such as combustion of fossil fuels, nuclear reactors, or falling water. The development of the United States economy has been closely linked to the use of electricity. Demand for electricity has increased much faster than the national economy. In the early 1970s, an interruption in the flow of petroleum products required conservation of energy used to create electricity. Since that time, the use of other forms of energy has decreased. However, demand for electricity has grown faster than the national economy by 25% to 50%.

Of the energy used in the United States, approximately one third is used in the generation of electricity. See Figure 2-1. Approximately 40% of the electricity generated is used in industry, 34% is used in residences, and 26% is used in commercial applications. Prior to 1970, less than 8% of residences were heated electrically. With technological advances in insulation and building materials, the energy required for heating a residence has decreased. This has resulted in an increase of new dwellings built with electric heat. Currently, approximately 50% of all new dwellings are heated electrically.

In 1973, fossil fuels provided 35% of the electricity nationally. Increased demand for electricity and petroleum products required a greater reliance on coal and nuclear energy. Since 1973, falling water (hydroelectric) power plant's share of generating electricity has dropped from 16% to 11%. Coal usage has grown to 57%, and nuclear energy has grown from 4% to greater than 16% for generating electricity. The increased demand for electricity has resulted in a greater reliance on nuclear power for generating electricity. For example, recent studies show that 86.5% of all Commonwealth Edison customers in northern Illinois use electricity generated by nuclear power plants.

In fossil fuel power plants, the combustion of fuel oil, coal, or natural gas is used to produce heat. In nuclear power plants, nuclear fission in a nuclear reactor is used to produce heat. In both fossil fuel and nuclear power plants, generating electricity

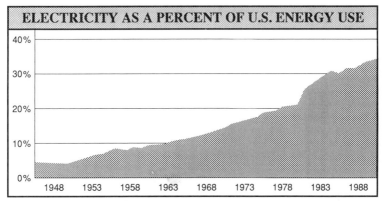

U.S. Council for Energy Awareness

Figure 2-1. Approximately one third of the energy used in the United States is used to generate electricity.

starts with the production of heat. Heat is used to convert water to steam. The steam produced is used to drive steam turbines. Steam turbines drive generators to produce electricity. As the steam leaves the steam turbine, it is condensed to water and returned to the boiler.

Use of water as an energy source for generating electricity (hydroelectric power) limits the ability to generate electricity to areas with falling water. The supply of fossil fuels is subject to interruption from other countries and is limited in supply. The construction of a nuclear power plant is more costly than a fossil fuel power plant. However, there are many advantages to a nuclear power plant.

Uranium 235 (^{235}U) is a plentiful commodity compared with fossil fuels. Nuclear power plants are not limited to sites with large land requirements for fuel storage. In addition, adverse weather conditions do not affect the storage of fuel or the operation of the plant. Nuclear power plants have no large fuel storage requirements and can be built anywhere as required. In addition, there are increasing concerns regarding acid rain and ash removal from burning coal. Nuclear power plants are classified as either stationary or portable.

Stationary Nuclear Power Plants

Stationary nuclear power plants are plants constructed in a permanent location. These plants are normally located near a river or lake for an adequate supply of cooling water. In some cases, cooling ponds or lakes are constructed for this purpose. In addition, stationary nuclear power plants are located for efficient distribution of electricity generated. See Figure 2-2.

U.S. Council for Energy Awareness

Figure 2-2. Stationary nuclear power plants are constructed near a source of water.

Portable Nuclear Power Plants

Portable nuclear power plants are plants designed for usage in any location. These plants are very useful in emergency situations and for providing electricity to remote geographic locations. Portable nuclear power plants eliminate problems of transportation of fuel to the point of use. A portable nuclear power plant was used in the early 1960s to successfully provide electricity to a research station beneath the ice cap in Greenland. This specially designed nuclear reactor was constructed so that it could be dismantled and transported by ship, truck, plane, and snowmobile to the point of use. See Figure 2-3.

Figure 2-3. Portable nuclear power plants can be used in a variety of locations.

Naval Vessels. The first nuclear-powered naval vessel was the United States submarine *Nautilus* in 1955. See Figure 2-4. The success of the USS *Nautilus* and later submarines proved the benefits of nuclear-powered submarines. Nuclear-powered submarines can maintain normal operations for over two years without refueling.

Nonnuclear-powered submarines are operated underwater using electricity from batteries. This requires the submarine to surface frequently to charge the batteries. The batteries are charged by diesel-powered generators. Nuclear-powered submarines can stay submerged longer because a minimal amount of fuel is required. In addition, the fuel requires less space than diesel fuel and related equipment, which can be used for other purposes. The USS *Nautilus* pioneered the use of nuclear power for submarines. At the present time all but two of the submarines in the U.S. Navy are nuclear-powered. No diesel-powered submarines are planned.

Figure 2-4. Nuclear-powered submarines can stay submerged longer than diesel-powered submarines.

Another major step in the development of nuclear power plants was the completion of the United States Nuclear Ship *Savannah* in 1959. The *Savannah*, a 22,000 ton merchant ship, was designed as an experimental ship to research the problems of nuclear-powered ships. The experiment was successful, allowing the ship to carry 10,000 tons of cargo and cruise 300,000 nautical miles on one fueling. Approximately 35 pounds of ^{235}U were required to cruise 90,000 miles. Since the *Savannah*, many commercial and military nuclear-powered naval vessels have been built by the United States and other countries around the world. See Figure 2-5.

RADIOISOTOPES

Radioisotopes are atoms that emit radiation. Many radioisotopes occur naturally in nature, but many more can be created artificially. Radioisotopes have replaced many of the uses of radiation produced by large, expensive X-ray machines. Radioisotopes are identified by their chemical name and mass number. For example, the radioisotope carbon 14 (^{14}C) has the chemical properties of

Figure 2-5. Nuclear-powered ships can make extended cruises on one fueling.

carbon 12 (^{12}C) except that ^{14}C consists of two additional neutrons, which makes it radioactive.

Radioisotopes emit radiation in the form of rays or particles. These rays can penetrate various substances such as concrete, body tissue, or paper. The amount of radiation that penetrates depends on the thickness, density, type of radiation, and chemical makeup of the material affected. See Figure 2-6.

Radioisotopes have radioactive lives that last a specific amount of time. Radioisotopes lose their radioactivity over time through radioactive decay. *Radioactive decay* is the loss of radioactivity

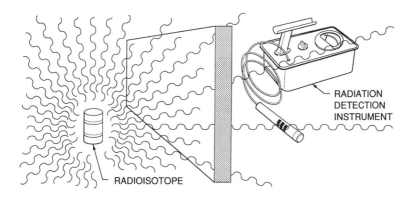

Figure 2-6. Radioisotopes emit radiation in the form of rays or particles.

by radioisotopes over a period of time, which is measured in half-lives. The *half-life* of a radioactive material is the time required for a quantity of that material to lose one half its radioactivity. During this process, half of the atoms decay into another chemical element or isotope. Half-lives of radioisotopes vary from fractions of a second to millions of years. The half-life of a radioactive material can be predicted accurately.

Radioisotopes of a specific chemical element have the same chemical properties as nonradioactive atoms of the same element. Radioisotopes that have the same chemical properties as nonradioactive chemical elements are available for many uses. This allows a specific chemical element to be traced by adding a radioisotope to the specific chemical element. *Radiotracers* are radioisotopes used for tracing specific chemical elements.

Because of these characteristics, radioisotopes can be introduced into humans, animals, plants, or machines and their movements can be followed by radiation detection instruments. For example, if radiotracers are introduced into the water of a pipeline under pressure, very small leaks not found by ordinary pressure tests can be detected. See Figure 2-7.

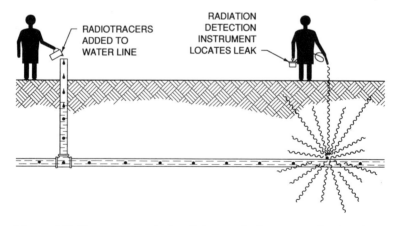

Figure 2-7. The movement of radiotracers in humans, animals, plants, and machines can be followed by radiation detection instruments.

Medical Applications

Millions of medical procedures requiring the use of radioisotopes are performed each year. Additionally, 30% of all biomedical research depends on the use of radiotracers. While the amount of radioactivity in these applications is small, an increased sensitivity to the potential effects of radioactive particles has lead to the development of imaging devices that enable doctors to watch vital organs function, identify growths and blockages, and detect early signals of diseases without exploratory surgery.

Approximately 50% of all patients admitted to hospitals in the United States are diagnosed using radioisotopes. In medical diagnostic work, radioisotopes allow doctors to obtain information about the internal body without the use of X-ray equipment. This reduces exposure to radiation from X rays required for diagnostic work. The X ray was the first device to allow an internal view of the body. Radiation produced by X-ray machines and by the naturally radioactive chemical element radium 226 (^{226}Ra) has been used for years in medical applications. The use of radioisotopes, radio waves, magnetism, and sound allow doctors to track the flow of body fluids; scan internal organs, bones, and other matter; and perform intricate operations that were not possible in the past.

Magnetic resonance imaging (MRI) uses radio waves and a strong magnetic field to scan the body. X rays are not involved. MRI momentarily alters the alignment of hydrogen protons in the body and reflected sound waves are gathered to form the image. See Figure 2-8.

Sonography (SONO) works with high-frequency sound waves to create images. A crystal converts electric pulses into sound waves that penetrate the body. The time delay in the returning sound waves are converted by a computer into a video image. Radioisotopes are not required for SONO.

Digital subtraction angiography (DSA) uses substances opaque to X rays coupled with computers to generate images. DSA is particularly useful in showing blocked arteries.

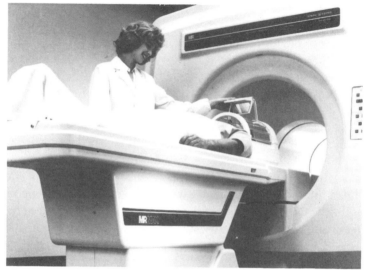

Figure 2-8. Magnetic resonance imaging (MRI) is used to scan internal portions of the body. X rays are not involved.

One of the most important radioisotopes used for this purpose is iodine 131 (^{131}I). Doctors use ^{131}I radiotracers to determine blood output of the heart, locate brain tumors, and study the functioning of the thyroid gland.

To trace the flow of blood in a patient, radioisotope sodium 24 (^{24}Na) is injected into the bloodstream. This allows the flow of blood to be traced throughout the various parts of the body. Different radioisotopes collect in concentrations in certain parts of the body. Arsenic 74 (^{74}As) is used to locate brain tumors. Brain tumors collect ^{74}As, and radiation monitors placed on the head indicate the position of the tumor. X rays are not sensitive enough to perform this task. See Figure 2-9.

Single photon emission computed tomography (SPECT) is the process of imaging an internal organ using a radioisotope. In SPECT, radioisotopes are administered and concentrate in a specific organ. Using a camera that senses radiation emitted from the organ, an image is created. The image is then analyzed for proper function and abnormalities.

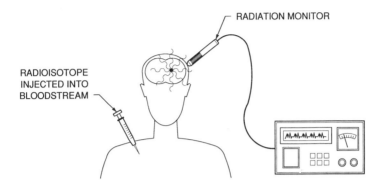

Figure 2-9. Radioisotopes are used to locate brain tumors.

Soon after the discovery of X rays and radium, just before 1900, doctors learned that radiation has the power to destroy cancer cells. Radioisotopes and radiation treatment have improved cancer treatment by destroying specific tissue with less damage to surrounding tissue. Radioisotopes have also provided additional treatment for specialized types of cancer.

Scientists are researching radioactive substances that can be absorbed only by cancer cells. If such a substance were introduced into the body, the cancer cells would be identified for a large dose of radiation with little danger to the other parts of the body.

Scanners are detectors that show radiation introduced into the body. Large areas of the body can be checked for tumors and other abnormalities using scanners. See Figure 2-10.

Equipment Sterilization. Radiation is also used to sterilize medical equipment and supplies such as scissors, needles, bandages, and sponges. The sterilization procedure using radiation is performed with the protective wrapping or containers holding the equipment intact. Sterilization of medical equipment by radiation replaces the old toxic gas method, eliminating the potential for toxic gas release. See Figure 2-11.

National Institutes of Health

Figure 2-10. Scanners are detectors that show radiation in the body.

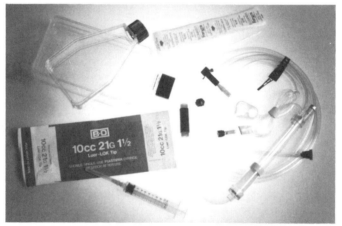

Radiation Dynamics, Inc.

Figure 2-11. Radiation is used to sterilize medical equipment and supplies.

Science Applications

Radioisotopes have provided a vehicle for research and the development of new techniques in the field of science. The ability to tag atoms with a specific radioisotope allows specific characteristics of a scientific process to be identified and traced over the years. Radioisotopes have enabled scientists to determine the age of the Earth. They have also provided many clues related to human and animal development. Other uses of radioisotopes in science range from authenticating a historic painting to tracking down an industrial polluter.

Date Analysis. *Date analysis* is the process of measuring the age of archaeological finds. The two most common methods of date analysis are radiocarbon dating and fission track dating. *Radiocarbon dating* is a date analysis method that measures radioactive carbon 14 (^{14}C). Carbon 14 is constantly being produced in the Earth's upper atmosphere by irradiation of nitrogen atoms in the air from neutrons from the cosmic rays. *Irradiation* is exposure to radiation such as X rays or a stream of neutrons. These ^{14}C atoms are present in all living things and decay with a 5700 year half-life. When a plant or animal dies, no new ^{14}C atoms are taken into that life system. Scientists are able to measure the amount of ^{14}C left in any sample and compare it to the amount of ^{14}C present in a current sample of similar material. These samples may be materials such as bone, teeth, wood, etc. The scientists then calculate how many half-lives have passed since the material stopped acquiring ^{14}C in its living state. Material age up to 60,000 years can be determined using this method. The margin of error is within approximately 100 years. See Figure 2-12.

The half-life of ^{14}C is 5700 years. This limits accurate dating to 60,000 years or less. For materials older than 60,000 years, scientists use longer atomic clocks such as the decay of uranium 238 (^{238}U) into lead 206 (^{206}Pb), or potassium 40 (^{40}K) decaying into argon 40 (^{40}Ar), which have longer half-lives. The samples are analyzed using sensitive measuring equipment and the fission track dating method.

Fission track dating is a date analysis method that records fission tracks on a film placed next to the material being dated. The fission tracks are counted to determine the amount of ^{238}U or ^{40}K left in the material. As in the radiocarbon dating method, the amount of each chemical element found is compared with a

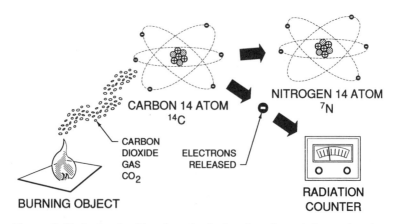

Figure 2-12. In the traditional method of radiocarbon dating, a burning object releases CO_2 gas. Carbon 14 in the gas releases electrons as it changes to ^7N. The number of electrons released determines radiocarbon content of the object.

current sample of similar material. Using fission track dating, a human skull found in Africa was determined to be two million years old. Fission track dating of ore found in uranium mines has established the age of the Earth at 4.5 billion years.

Neutron Activation Analysis. *Neutron activation analysis* is a detection process that can determine trace elements as small as one billionth of a part. The process involves placing a nonradioactive material in a nuclear reactor and bombarding it with neutrons. During the process, the proton-neutron ratio of the material is changed to an unstable condition. The material becomes radioactive and emits a characteristic gamma ray that is unique to the particular material. By using special gamma ray detectors, these radioactive elements can be identified in extremely low concentrations (trace amounts). Trace amounts this low are undetectable by standard chemical tests. See Figure 2-13.

Neutron activation analysis has become a valuable tool in law enforcement investigations. For example, the material to be analyzed could include a speck of soil on a suspect's shoe, a sample of polluted air, a sample of illegal waste discharge, or a human hair found on the suspect's clothes. The materials that can serve

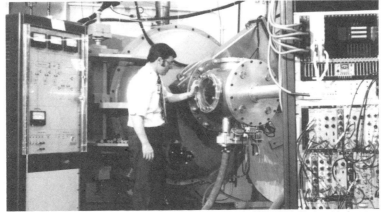

Oak Ridge National Laboratory

Figure 2-13. Material to be analyzed is bombarded with neutrons in neutron activation analysis.

as evidence are irradiated. Gamma ray emissions are recorded and serve as a fingerprint for identifying a specific material. Evidence from two sources can be compared and a positive identification can be made. Evidence obtained through neutron activation analysis has been used successfully in court.

Space Program. In the early years of the space program, electricity was obtained from lightweight batteries, fuel cells, and solar cells. As missions became more complex and travel distances increased, the need arose for more compact nuclear generators to produce electricity for a longer period of time. Nuclear generators powered by radioisotopes such as plutonium 238 (^{238}Pu) and strontium 90 (^{90}Sr) have been used for this purpose. Small portable nuclear reactors have been used in the space program to provide energy since the early 1960s.

By collecting the electrons given off by a radioactive material, the generator will produce electricity as long as the radioactive material lasts. The electric current produced is usually very small, but the generator can operate under conditions of vacuum and high temperature that could ruin batteries or other sources of electricity. See Figure 2-14.

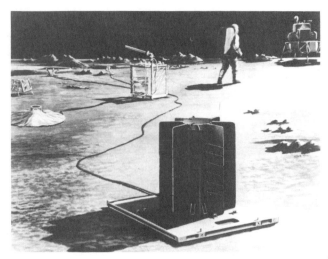

NASA

Figure 2-14. Nuclear-powered generators are used in the space program.

Industrial Applications

Nuclear energy is used widely for manufacturing and quality control. For example, if a radioisotope is placed on one side or inside of a metal casting or forging and a photographic plate is placed on the other side or outside, more rays will be stopped by the thick sections of metal than the thin sections. The photographic plate is darkened to different degrees depending on how much radiation passes through. A black-and-white shadow image of the object shows flaws and weak spots. Radiography is widely used for making nondestructive materials analysis on castings, forgings, and pipe, ship, and tank welds. See Figure 2-15.

Sometimes X rays cannot be used effectively when measuring the contents inside metallic materials. For example, it would be impossible to measure a small amount of explosive material in a steel casing. The steel would absorb too many of the X rays and details of the explosive material would be lost. This problem was encountered by the engineers in the early days of the Apollo Space Program. Hundreds of explosive couplers had to be tested

Figure 2-15. Radiography is used in quality control applications in industry.

to ensure proper operation. The engineers used a beam of neutrons to reveal defects that X-ray examination had missed. Neutrons pass through steel quite easily but are scattered or absorbed out of the beam by light hydrogenous materials. This allows metal parts to disappear, and the explosive material to show during inspection.

Gamma ray sources also are used as a nondestructive testing tool. They can be used instead of X rays to radiograph heavy metal pieces such as castings and thick-wall pipe welds. Gamma ray sources are more portable than X-ray machines, which tend to be more bulky. This allows better usage in examining welds in the field. These testing devices are used extensively in nuclear facility construction to examine pipe welds and structural supports.

Material Thickness. Some industries require accuracy of material thicknesses to one millionth of an inch (.000001″). This minute amount is a close tolerance requiring sensitive production and inspection equipment and processes. Maintaining the thickness of a sheet of material like this page (approximately .002″) is a more common task. The old method of measuring the thickness of paper during the manufacturing process involved stopping the production line, measuring the sample, and adjusting the rollers to the proper tolerance. This process limited productivity and was a time-consuming process.

An isotopic thickness gauge eliminates some of these problems by continuously monitoring the thickness of the material being produced. Rollers can be adjusted while the production line is operating to maintain the proper tolerance. The isotopic thickness gauge functions by placing a radioactive source on one side of the material to be measured. A radiation detector is placed on the other side of the material. The amount of radiation passing through the material to the detector is proportional to the thickness of the material. The detector meter can be calibrated in thousandths of an inch for a direct readout. See Figure 2-16.

A similar isotopic gauge can be used to measure the thickness of a material being plated on a base metal. In this application, the amount of material plated on the base metal determines the amount of radiation scattered back to the detector.

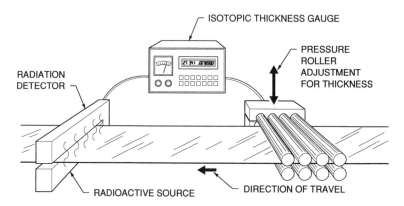

Figure 2-16. Isotopic thickness gauges use radiation to maintain material thickness.

Another example of testing for thickness involves the process of irradiating piston rings with neutrons to determine piston wear. The radioactive piston rings are installed in an engine and run for several minutes. Lubricating oil is pumped through the engine and past a radiation detector. The radiation detector determines the number of radioactive particles that have worn off the rings and are transported by the lubricating oil. Piston wear is then determined from this data.

Certain plastics such as polyethylene become tougher and more resistant to chemical attack after being irradiated. This permits thinner materials to be substituted for weight and material savings.

Static Eliminators. In dry weather, cloth, paper, plastics, and other nonconducting materials become electrically charged by friction when passing between rollers in manufacturing operations. The charge can be removed by exposing the material to radiation, which produces ions in the air to neutralize the charge.

Nuclear Explosives. Nuclear explosives have been used to successfully tap natural gas formations. The same power used for warfare can be isolated and used to obtain natural gas from deposits and other resources that would require extensive drilling operations. Many natural gas formations can be tapped using normal drilling techniques. In other very large natural gas formations, the pores of the gas-bearing rock are sealed with clay or other cementitious materials. This prevents gas from flowing through to a hole drilled using traditional methods.

It was theorized that if a large enough explosion could be set off underground, the rock would be shattered, which would release the natural gas for drilling using traditional methods. Project Gasbuggy was initiated in 1967 by drilling a well to receive a 26 kiloton nuclear device. This is equivalent to 26,000 tons of dynamite. After the device was installed, the well was sealed with concrete, sand, and a special expanding plastic to prevent leakage of any radioactive products into the atmosphere. The explosion vaporized nearby rock in all directions, allowing a test well to be drilled. The well yielded 135 million cubic feet of natural gas. Additional tests have been performed in Colorado using this technology. The cost of nuclear explosives has prohibited further development and usage of the technology. However, it has been predicted that this process will be used if increasing oil prices justify the expense of reaching hard to access fossil fuel reserves. See Figure 2-17.

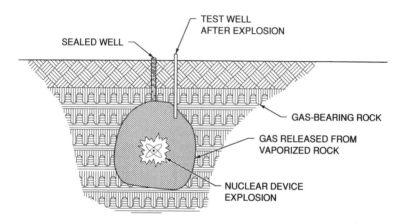

Figure 2-17. Nuclear explosives are used in petroleum exploration.

Petroleum Industry. Radiotracers are used in large oil refineries to pinpoint the location of a stoppage in a pipeline. Personnel using radiation detection instruments can find the exact location where the radiotracer source stops moving in the pipe. This provides quicker and less expensive troubleshooting than other methods to locate stoppage. See Figure 2-18. Oil refineries also use radiotracers to track various petrochemical products passing through the various processing stages. This provides continuous monitoring for product consistency and purity without stopping production.

Radioisotopes are used to analyze oil well drilling operations. During oil well drilling, a radioactive source is lowered into the drill hole along with a radiation detector. Scientists then measure

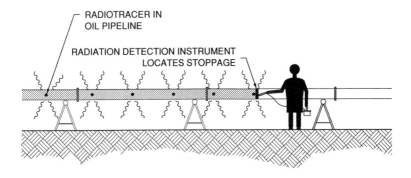

Figure 2-18. Radiotracers are used to locate stoppages in pipelines.

the radiation scattered back into the hole from the rock walls as the radiation source passes through the various rock layers. The information obtained indicates the presence or absence of crude oil in the surrounding rock.

Smoke Detectors. Smoke detectors placed in industrial buildings have been very successful in preventing injury to workers during a fire. The use of smoke detectors lowers insurance rates and provides a guarded environment for workers, equipment, and material. Smoke detectors are now commonly required in dwellings by many municipalities.

Detection of smoke in the air is accomplished by a sensor. A small amount of the radioisotope americium 241 (^{241}Am) is placed in an ionization chamber where an electric circuit is created using a battery supply. Smoke impedes the flow of electric current through the ionized air and breaks the electric circuit to trigger the alarm. See Figure 2-19.

Agricultural Applications

Radioisotopes are used to increase productivity of agricultural products. For example, radiotracers can be applied to pollen to track the pollenization process. The proper application method

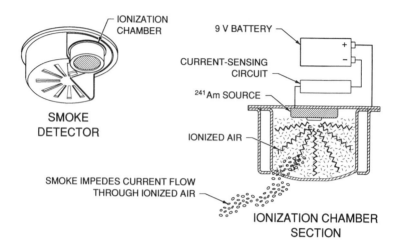

Figure 2-19. Smoke detectors are placed in industrial, commercial, and residential buildings.

of fertilizer can result in tremendous savings. Radioisotopes are used to determine the most efficient method to use when applying a fertilizer. For example, zinc is a vital nutrient to coffee trees. Radioisotope zinc 15 (^{15}Zn) was applied to coffee trees to test the effectiveness of various methods of fertilizer application. See Figure 2-20.

This produced the following results:
- Zinc applied to the soil resulted in 5% absorption by the plant.
- Zinc applied to the upper surfaces of the leaves resulted in 12% absorption.
- Zinc applied to the lower surfaces of the leaves resulted in 42% absorption in the plant.

The same techniques are used to determine the most effective method to apply insecticides and weed killers. Radioactive insecticides are sprayed on plants. The radioactivity emitted from the dead insects is then measured. Sterilization of male insects can result in a high proportion of sterile eggs being laid by the female and subsequent disappearance of the species. Insect infestations such as the Mediterranean fruit fly can be controlled

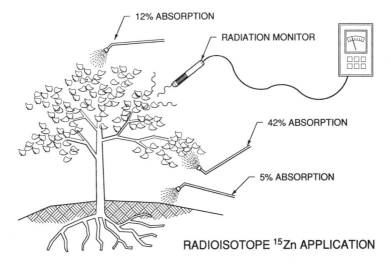

12% ABSORPTION

RADIATION MONITOR

42% ABSORPTION

5% ABSORPTION

RADIOISOTOPE ¹⁵Zn APPLICATION

Figure 2-20. Radioisotopes are used to determine the efficiency of fertilization application techniques.

by sterilization of the flies by low doses of gamma rays in portable traps.

Genetic Modifications. Genetic modification uses radioisotopes for agriculture and animal breeding. Radioisotopes are used as radiotracers for studying the reproductive processes that occur in plants and animals. This information and radiation experimentation have been used to produce better strains of plants and animals. In nature, most plants grown from seeds are like the plants that produced the seeds and most animals are similar to their parents, but a few are different. Such differences in the offspring are mutants.

A *mutant* is a significant and basic physical change to the chromosomes or genes of a plant or animal. Many mutants are worthless but some have much better traits than the original plant or animal. The best ones are developed in order to obtain new plants or animals that will produce more and better products. The best plants and animals are selected and used for breeding purposes. See Figure 2-21.

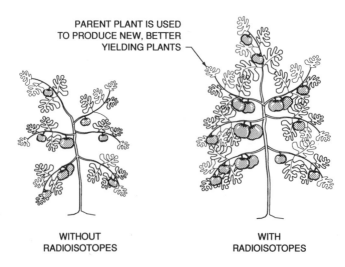

PARENT PLANT IS USED
TO PRODUCE NEW, BETTER
YIELDING PLANTS

WITHOUT
RADIOISOTOPES

WITH
RADIOISOTOPES

TOMATO PLANTS

Figure 2-21. Radioisotopes are used in genetic modification of plants and animals.

Another project to determine the effects of radiation on plants involved putting tomato seeds in space for six years. The National Aeronautics and Space Administration (NASA) placed millions of tomato seeds in containers built for maximum radiation exposure. The containers were aboard the Long Duration Exposure Facility satellite, which was retrieved by the space shuttle Columbia in early 1990.

Kits containing 50 space (flight) seeds and 50 regular (control) seeds are to be sent to science classrooms from the fifth grade to college level. Students will raise the seeds and compare germination rates, germination time, seed embryo, seed vigor, and fruit produced. Results will be sent to NASA for their report on the effects of seeds exposed to radiation in space.

If plants and animals are exposed to radiation the number of mutations can be greatly increased. Selecting and developing superior strains is a very long process but radiation makes it possible to get results much faster. For example, a project to develop rust-disease-resistant oats was successful after only about a year and a half of experimenting. This might have taken 10 years or more if only natural mutations were available. Irradiated seeds of many plants are continually being developed and are available for experimentation.

Irradiation. Irradiation of food makes chemical additives and preservatives unnecessary by destroying bacteria, viruses, molds, and insects. In this process, food is exposed to gamma rays from the radioisotopes cobalt 60 (^{60}Co) and recently cesium 137 (^{137}Cs). In addition, irradiation also increases shelf life and inhibits the sprouting of vegetables. For example, potatoes are less likely to develop sprouts when irradiated. In one experiment, potatoes that were not irradiated showed substantial sprout growth over 16 months. Potatoes that received the most radiation showed no sprout growth. See Figure 2-22.

Irradiation requires less energy than other preservation methods. It does not produce radioactivity in the food to which it is

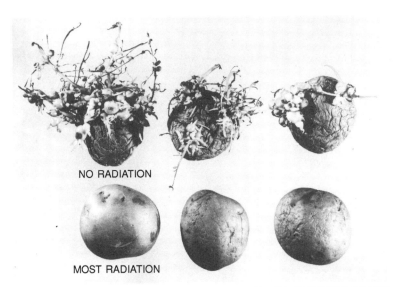

NO RADIATION

MOST RADIATION

Brookhaven National Laboratory

Figure 2-22. Irradiation of food makes chemical additives and preservatives unnecessary.

applied and the food is perfectly safe for human consumption. Irradiation extends the life of the food product but does not preserve it entirely. However, this is important since in the United States, the loss on orchard products alone is near 100 million dollars each year. Irradiation also eliminates the need to use potentially dangerous chemicals as preservatives. High-level radiation will preserve food without refrigeration or freezing for many years. This is particularly important in certain Third World countries where extensive refrigeration is not available. Thus, fruit that has been irradiated can be exported to these countries, which increases the market for these crops. The Food and Drug Administration, which has the responsibility of approving processed food for human consumption, has approved the radiation sterilization process for specific foods. See Figure 2-23.

The U.S. Army maintains a radiation laboratory at Natick, Massachusetts, for research in food irradiation. In one year, the U.S. Army produced 30,000 pounds of radiation-sterilized bacon

FOODS APPROVED BY FDA FOR IRRADIATION TREATMENT		
Food	**Purpose**	**Date Approved**
Fruits and vegetables	To slow growth and ripening and to control insects	April 1986
Dry or dehydrated herbs, spices, seeds, teas, vegetable seasonings	To kill insects and control microorganisms	April 1986
Pork	To control *Trichinella spiralis* (the parasite that causes trichinosis)	July 1985
White potatoes	To inhibit sprout development	August 1964
Wheat, wheat flour	To control insects	August 1963

Food and Drug Administration

Figure 2-23. The Food and Drug Administration has approved the use of irradiation for specific foods.

for its own use. In the same year, 40,000 pounds of potatoes were radiation-sterilized for use on military bases.

SKILLED TRADES

Nuclear energy has provided many new job opportunities for skilled tradesworkers because of the need to fabricate special parts and equipment for experimental work, the custom manufacturing of many items, the precise tolerances that must be maintained, and the great amount of complex machinery and equipment that must be serviced. See Figure 2-24.

Duke Power Co.

Figure 2-24. Skilled tradesworkers construct and maintain nuclear power plants.

In laboratories, hospitals, plants, and other establishments that use radioisotopes, most employees do the same work as they would otherwise. The difference is that almost all such work is more complicated because of the safety precautions that must be taken.

Journeymen from many skilled trades are employed in all parts of the country in installations under the supervision of the Nuclear Regulatory Commission (NRC), and the number continues to increase. See Figure 2-25. This work includes construction and maintenance of laboratories and other buildings, reactors, electrical power generating plants, construction and maintenance of electrical safety and monitoring devices, and the manufacture of reactor components and instruments. Additionally, skilled tradesworkers are required for decontamination and decommissioning work. Journeymen from almost every skilled trade are employed in this work, including electricians, plumbers, pipefitters, steamfitters, welders, machinists, carpenters, and sheet metal workers. It has been estimated that approximately one third of the cost of a nuclear power plant is required for plant instrumentation and control.

Nuclear Technicians

Nuclear technicians of all kinds are employed as radiographers, drafters, computer programmers, reactor and accelerator operators, radioisotope production operators, radiation monitors, laboratory technicians, decontamination workers, and waste treatment and waste disposal operators.

Electrical

Electricians perform the same type of work as those in other building construction and maintenance areas except for the additional requirements for safety. Monitoring and alarm systems, back-up systems in case of power failure, and many remote control and viewing systems require highly skilled and expert journeyman electricians.

Sheet Metal

Sheet metal workers install ventilation and exhaust systems, which play a large part in radiation safety. Radioactive particles from processes in the plant attach to dust particles and become

Duke Power Co.

Figure 2-25. Journeymen from many trades are required to set a nuclear reactor vessel in place.

airborne. All ventilation air must be filtered to prevent the escape of the contaminated dust particles into the outside environment. These filters and the associated piping become contaminated and require special handling during repairs and maintenance to prevent the spread of radioactive contamination to operating personnel and the general public.

In laboratories, each hood is connected to an outside exhaust and each exhaust line has a double-filtering system. A special filter is installed at the head to filter out radioactive particles. A second filter is installed on the blower to filter out any remaining particles. Special monitoring systems are installed above the roof line of the duct to determine if any radioactive materials are escaping into the exhaust air.

Plumbing

Safety precautions in nuclear power plants make plumbing work more complicated. In many installations, special plumbing systems are installed to prevent radioactive materials in the water from being carried down into the sewer system.

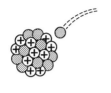

Nuclear Fission

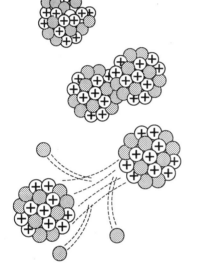

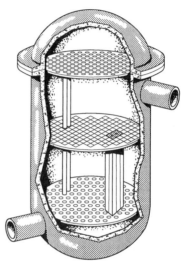

All nuclear power plants produce nuclear energy by the nuclear fission process, which is the splitting of atoms. Nuclear fission is maintained by the chain reaction that occurs when the neutrons from a split atom strike other atoms, causing them to split. The chain reaction is sustained in a nuclear power plant as fuel is fed and reaction conditions are controlled. The most commonly used fuel is uranium 235 (^{235}U).

The major parts of a nuclear reactor are the fuel, moderators, control rods, reactor vessel, coolant, and shielding. The most common type of nuclear reactor used in the United States is the light water reactor (LWR). The two types of light water reactors are the boiling water reactor (BWR) and the pressurized water reactor (PWR).

NUCLEAR REACTIONS

Nuclear reactions occur continually in nature. Nuclear fission is the most common nuclear reaction. It is used to create heat for production of steam in nuclear power plants. Great amounts of heat are produced by the splitting (fissioning) of the atom. Steam produced from boiling water from the heat is used to drive steam turbines, which drive generators for the production of electricity. Since 1942, when the first pile was built, nuclear reactors have been constructed all over the world. Power production capability is crucial to the United States economy. In 1989, the status of nuclear power plants in the United States was

- Operating licenses 110
- Construction permits 13
- Plants on order 2

Total 125

Nuclear power plants in the United States generate 116,100 megawatts (MW) of electricity for approximately 20% of the total power production in the United States. See Figure 3-1.

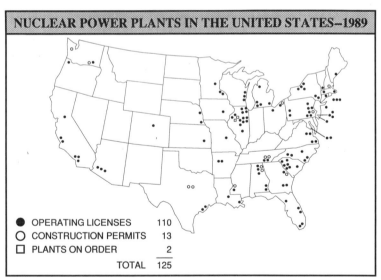

Washington Public Power Supply System

Figure 3-1. Nuclear power plants have increased their percentage of total power production to meet the demands of the U.S. economy.

A comparison of electricity generated from a nuclear power plant and electricity generated from a generator at a large dam shows how much electricity is generated using the fission process. The current generation of nuclear reactors rated at 1100 MW with about 65% to 80% operating time yields 700 MW to 800 MW per year. The Grand Coulee Dam, located in Washington state, is the largest dam in the United States at approximately one mile long. The dam is rated at 6300 MW but must divert water flow to meet irrigation needs and flood control. This results in the Grand Coulee Dam producing an average of 2200 MW per year. One nuclear power plant can produce as much as one third of the annual output of the Grand Coulee Dam.

Presently, Illinois has the largest number of operating plants (13) with two on order. Pennsylvania has eight operating plants with one construction permit issued. The majority of the nuclear power plants are located east of the Mississippi River.

Nuclear fusion is the source of energy of the sun and other stars. Light, heat, and other forms of energy are released when hydrogen atoms in the sun combine and become helium atoms. In some stars hotter than the sun, helium atoms combine to form carbon. In stars with more heat, carbon atoms combine to form still heavier atoms. This results in the production of chemical elements through a combination of fusion, fission, and radioactivity. These reactions occur constantly on stars throughout the universe. See Figure 3-2.

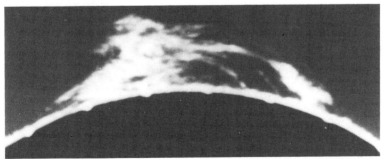

Mount Wilson and Palomar Observatories

Figure 3-2. Fusion reactions occur continually on the sun and other stars.

Nuclear Fission Process

Nuclear fission starts with the uranium atom. Inside the uranium atom there is a large number of protons. The number of protons in a nucleus determines the specific chemical characteristics of the atom. All hydrogen atoms have one proton, carbon has six protons, etc. The uranium (^{235}U) nucleus is the easiest natural nucleus to split because of its large number of neutrons. It contains 92 protons and 143 neutrons. See Figure 3-3.

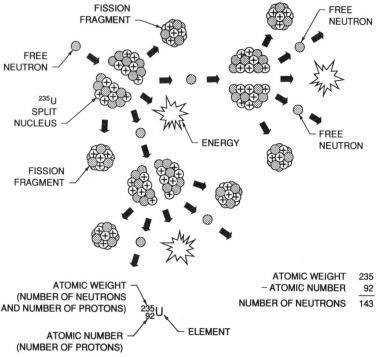

Figure 3-3. The ^{235}U atom is the easiest to split because of the large number of neutrons present in the nucleus.

The number of neutrons in the nucleus is determined by subtracting the number of protons (atomic number) from the number of neutrons and protons (atomic weight), identified by the mass number. Circling around the nucleus are the electrons.

Normally there is the same number of electrons as protons in an atom. An excess of either protons or electrons will result in the atom having a positive or negative electrical charge. Atoms with a positive or negative electrical charge are ionized. An *ion* is an atom with a positive or negative electrical charge. Protons have a like positive charge. This normally would cause protons to repel each other. However, powerful nuclear forces cause the protons to remain together in the nucleus.

Binding Energy. *Binding energy* is the energy required to overcome an atom's nuclear forces to release protons and neutrons. Binding energy takes up weight in the nucleus of an atom. For example, a helium nucleus, consisting of two protons and two neutrons, weighs more than the protons and neutrons weigh separately outside the nucleus. When separated, the binding energy that holds the helium nucleus together is released. This causes a change in the mass. See Figure 3-4.

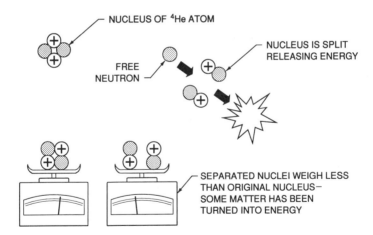

Figure 3-4. Binding energy accounts for a portion of the weight of the nucleus of an atom.

Accelerators. In the 1930s, scientists experimented with bombarding atoms in accelerators. An *accelerator* is a device used to smash atoms by shooting a stream of high-speed particles at large atoms such as uranium. Using an accelerator, it is possible to overcome nuclear forces in large atoms such as uranium, causing them to split apart or fission. The fission process occurs when a free neutron enters the nucleus of a fissionable atom. The nucleus immediately becomes unstable, vibrates, and splits into two fission fragments, forming two nuclei of lighter atoms and emitting an average of 2.7 free neutrons. Since the binding energy needed to keep these lighter nuclei together is less than the binding energy of the original ^{235}U atom, energy is released. This energy will not turn back into mass again. It remains energy and is released in the form of heat caused by the kinetic energy (energy of motion) of fragments colliding with surrounding atoms and molecules and radiation. See Figure 3-5.

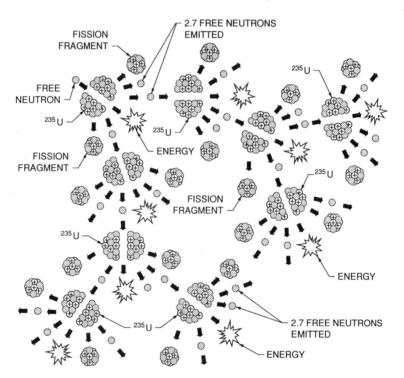

Figure 3-5. Mass turned into energy is created by the binding energy released during nuclear fission.

Einstein's Theory. The process of turning mass into energy was anticipated by Einstein in 1905. His formula $E = mc^2$ predicts that a small amount of mass (m) can be transformed into a large amount of energy (E), and that the amount of energy can be calculated by multiplying the mass times the speed of light squared (c^2). In addition to fission fragments and heat, a fissioning nucleus also frees two or three additional neutrons. Some of the free neutrons strike other fissionable atoms. The rate at which these free neutrons are emitted is the key to sustaining the nuclear reaction. Only atoms that have nuclei with more neutrons than protons can produce a sustained nuclear reaction.

Chain Reaction. During World War II, the efforts of the United States were focused on fissioning the uranium atom to release a great amount of energy for weapons development. The Germans were also working on the same project. The primary goal of the United States was to develop the atom bomb before Germany. However, before an atom bomb could be developed, it was necessary to prove that nuclear fission could continue automatically after it was started. Before 1942, scientists had been able to split the atom only under laboratory conditions. It took more energy to make the atoms split than the fission released and the process stopped immediately unless more energy was fed in.

The scientists theorized that if conditions were right, some of the neutrons that escaped from the nucleus when a uranium atom was split would hit other ^{235}U atoms and also cause them to fission. If an average of one neutron from every split atom struck and fissioned another atom, the reaction could continue indefinitely and stay under control. But if more than one neutron per fission hit and split another atom the reaction would go out of control, releasing sufficient energy to create an atomic explosion. A *chain reaction* is the process in which released neutrons from an atom strike and split other atoms, which repeats the procedure. In a chain reaction, one reaction leads to the next reaction, which causes the reaction to be sustained.

To control this potentially dangerous process, it was necessary to regulate neutron production to provide just one neutron per fission to be available to produce another fission. Approximately 2.7 neutrons per fission are produced, necessitating the removal of approximately 1.7 neutrons per fission from the reactor core.

Chain reaction in nuclear explosives occurs by the compounding number of neutrons coming in contact with one another. If two neutrons from a split atom hit two more atoms, these atoms split and release a total of four neutrons. The process continues to repeat itself very quickly. See Figure 3-6.

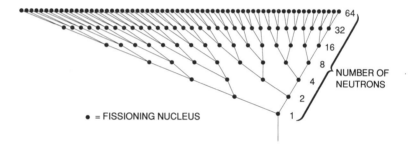

Figure 3-6. A nuclear chain reaction is sustained as free neutrons collide with other nuclei and cause fission. Fission results in the creation of more free neutrons.

The fissioning of one atom releases several neutrons. In controlled conditions, scientists can predict approximately how many neutrons will be in motion and how many atoms will be split. As long as each atom that fissions causes one other atom to split, the reaction will continue at the same power level but will not grow. If the power level is maintained, the reaction is critical. If the average number of atoms split is more than one, the power level will increase and the reaction is supercritical. If the average number of fissions caused by one atom splitting is less than one, the power level will decrease and the reaction will stop, creating a subcritical condition.

In a *subcritical reaction*, the average number of fission-producing neutrons is less than one. In a *critical reaction*, the average number of fission-producing neutrons is one. In a *super-critical reaction*, the average number of fission-producing neutrons is more than one.

Nuclear Fusion Process

The nuclear fusion process provides a great amount of energy from common chemical elements. However, the technology does not currently exist to allow any commercial applications of nuclear fusion. Nuclear fusion releases much more heat than fission and is easier to control. The hydrogen, water, or other hydrogen compounds that could be used are more plentiful than the uranium materials used in nuclear reactors. See Figure 3-7.

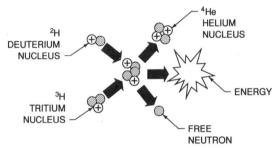

Figure 3-7. Nuclear fusion involving deuterium and tritium nuclei is still in the experimental stage.

NUCLEAR FISSION CONDITIONS

Nuclear fission conditions control the activity of neutrons in the fission process to maintain a chain reaction. This control is accomplished as a function of the quantity of neutrons, and the conditions created by the reactor parts. In addition, these conditions also have a bearing on the average number of neutrons present in the reactor. Many of the neutrons will escape through the surface of the reactor and not be involved in the fission process.

Critical Mass

To sustain a chain reaction in a nuclear reactor, the core must be designed to retain the required amount of neutrons. The *core* is the central part of the nuclear reactor that contains the fuel elements. The size of the core is a factor in maintaining the chain reaction. The core must be large enough to contain neutrons inside the reactor to sustain the chain reaction. *Critical mass* is the smallest mass of fissionable material that will support a self-sustaining chain reaction under certain conditions. If the core is too small, too many neutrons will escape, causing the chain reaction to cease. Increasing the size of the core to the critical mass makes a chain reaction possible.

Critical Mass Shape. In addition to the size of the critical mass, the shape also affects the nuclear reaction. A change in the shape of the critical mass can make the reaction subcritical. For example, a critical mass in the shape of a cube has the same amount of fissionable material as a critical mass in the shape of a plate. However, because the plate has less space for neutron reactions and more surface area, more neutrons escape the reaction. This causes the reaction to become subcritical. See Figure 3-8.

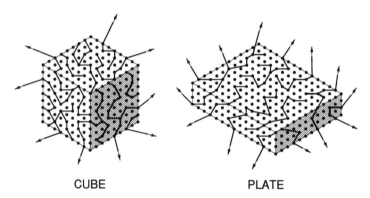

CUBE   PLATE

Figure 3-8. Critical mass in the shape of a cube allows fissions to occur, resulting in a chain reaction. The same critical mass in the shape of a plate reduces the fissions occurring, causing the chain reaction to cease.

Reflectors

A reactor core smaller than the critical mass can sustain a chain reaction if surrounded by a reflector. A *reflector* is a layer of material surrounding a core, which reflects neutrons that would otherwise escape back into the core. The reflected neutrons cause more fissions, which improve the efficiency and economy of the reactor. Reflectors commonly used include graphite, beryllium, and natural uranium.

Reaction Control

A nuclear reaction is controlled by the amount of fissions that occur. For example, in a nuclear explosion, the chain reaction is supercritical instantly. The chain reaction remains supercritical as long as possible. A nuclear reactor produces the same amount of heat and energy as a nuclear explosion. However, the energy and heat are monitored and released at a controlled rate.

An example of the difference between a nuclear explosion and a reaction in a reactor is the lighting of a firecracker. If the firecracker is lit at the fuse, ignition of all the powder occurs instantly. If the firecracker is opened and the loose powder lit, the powder is ignited slowly. The same amount of powder is consumed, but at a different rate. A nuclear reactor cannot explode like nuclear explosives. Fuel used in nuclear power plants commonly contains only 3% ^{235}U. Nuclear explosives require fissionable fuel in greater concentrations of ^{235}U than that used in nuclear power plants.

NUCLEAR REACTOR PARTS

Nuclear reactors supply the heat required to change water into steam in a nuclear power plant. In a nuclear reactor, a chain reaction occurs at a constant rate. The chain reaction is sustained as fuel is exposed, and reaction conditions are controlled. The six main parts of a reactor are the fuel, moderators, control rods, reactor vessel, coolant, and shielding. See Figure 3-9.

Fuel

The fuel is the heart of a nuclear reactor. The fuel used in most reactors consists of pellets of uranium encased in metal tubes

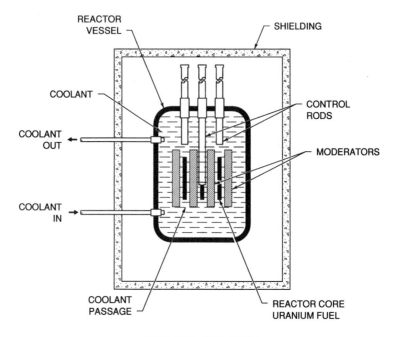

NUCLEAR REACTOR

Figure 3-9. Nuclear reactors are designed to create and sustain nuclear fission chain reaction safely and efficiently.

(rods). The pellets are less than ½" in diameter but each has as much energy as 120 gallons of oil. The fuel pellets are stacked in the rods, which are approximately 12' long. The metal tubes are bundled to form fuel assemblies. The fuel rods are spaced to permit coolant to flow between each rod. See Figure 3-10.

The most commonly used fuel is ^{235}U. Uranium occurs naturally and is 100 times more common than silver. However, ^{235}U is quite rare in relation to the total uranium mined. Approximately 0.7% of uranium mined is ^{235}U. The remaining 99.3% is ^{238}U. Before ^{238}U can be used in a nuclear power plant it must be enriched to a 3% concentration of ^{235}U.

The most common uranium isotope is ^{238}U but it is not fissionable under most conditions. However, it is fertile, which

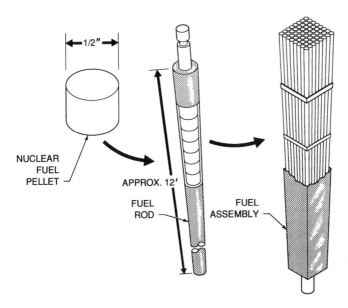

Figure 3-10. Fuel pellets are stacked in fuel rods. Spaces between the fuel rods provide adequate cooling.

indicates that it will readily absorb neutrons. Instead of fissioning, ^{238}U absorbs these neutrons and is converted to plutonium 239 (^{239}Pu). Plutonium 239 can be used in place of ^{235}U. This process allows the reaction to create heat and energy and produce fuel at the same time.

Natural uranium consists mainly of atoms of the isotope ^{238}U. These ^{238}U atoms will not split unless they are hit by fast-moving neutrons, and even then no chain reaction occurs. Just the opposite occurs with ^{235}U. Slow-moving (or low-energy) neutrons are absorbed and result in fission and a chain reaction can be sustained. See Figure 3-11.

One of the fuels used in reactors is enriched uranium, which is uranium that has been processed so that it contains many more atoms of the rare isotope ^{235}U than natural uranium. The other fuel is plutonium, a chemical element produced in reactors and accelerators.

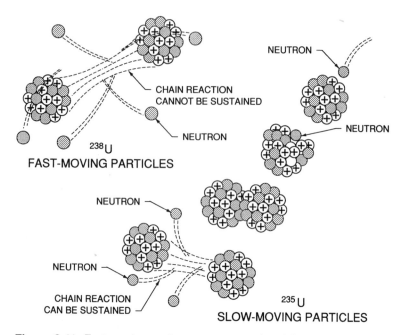

Figure 3-11. Fast-moving neutrons pass around and through nuclei without causing fission. Slow-moving neutrons fission nuclei and sustain the chain reaction.

Fuel Processing. Fuel used in a nuclear reactor is prepared from mined uranium ore. The uranium ore is crushed and milled at a mill into yellowcake. *Yellowcake* is processed uranium ore ready for additional processing and enrichment. It contains many varieties and isotopes of uranium. However, approximately 1% of the yellowcake is ^{235}U and the rest is ^{238}U. The amount of ^{235}U can be increased by converting the yellowcake to another chemical form and processing it at the enrichment plant. See Figure 3-12.

The enrichment plant increases the concentration of ^{235}U required for use in the reactor. Uranium contains 0.7% of ^{235}U in its natural state. It is enriched by converting it to its gaseous form, uranium hexafluoride (UF_6). The gas is then passed through small holes. The ^{235}U will pass through the holes more

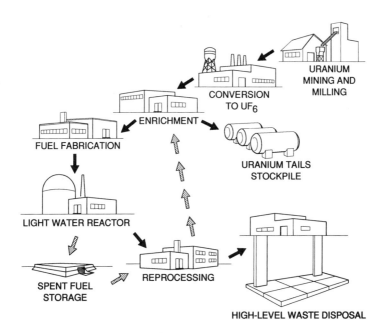

Figure 3-12. Uranium from the mine requires processing and enrichment prior to use in a reactor.

easily than ^{238}U, causing the two to separate. The enriched uranium is then converted back to its solid form (uranium oxide) for further processing.

In the enrichment process, natural uranium is divided into 3% enriched uranium and 0.2% tails. *Tails* is depleted material that is unusable in a nuclear reactor. Tails must be properly stored. The enriched uranium is fabricated into fuel elements. The fuel elements are shaped into pellets and grouped into fuel assemblies in the reactor. A fuel assembly will produce power for approximately three to five years before it is removed from the core. After the fuel is removed from the reactor, some uranium and plutonium can be salvaged for reuse. This fuel must be chemically reprocessed to produce enriched uranium. This process allows enriched uranium to be recycled many times in the reactor.

Moderators

Moderators are nuclear reactor parts that slow down neutrons in the fission process. The ^{235}U atom is fissioned with a slow-moving neutron that can be absorbed into the nucleus. Most neutrons released during fission travel too fast to cause fission in another ^{235}U atom. Neutrons freed during fission travel at approximately 12,000 miles, or greater, per second and pass by and through ^{235}U too quickly to cause a reaction. If these fast-moving neutrons pass through moderators such as water, heavy water, beryllium, or graphite, they collide and lose energy. Neutrons slowed to approximately 1 mile per second have a much better chance of causing fission when colliding with ^{235}U atoms. Water is an efficient moderator and also serves as a coolant. See Figure 3-13.

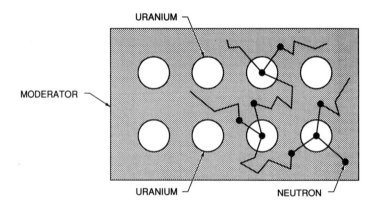

Figure 3-13. Water is the most common moderator used for nuclear reactors.

The effective use of moderators allowed the first fission chain reaction to be conducted. The first nuclear reactor was constructed of graphite blocks with uranium rods running through the blocks. The blocks were arranged so that the uranium rods formed a lattice completely enclosed by graphite. Fast-moving neutrons released by fission of uranium atoms collided with atoms in the graphite moderator. These collisions reduced the speed of the neutron enough so it could be absorbed by a ^{235}U atom in the fuel rod to continue the chain reaction.

Control Rods

Control rods control the rate of the chain reaction in a nuclear reactor. Control rods are interspersed among the fuel assembly. They are made from materials such as boron and cadmium, which absorb neutrons without fissioning. Boron is the most commonly used material for control rods. The position of the control rods regulates the speed of the chain reaction. See Figure 3-14. When the control rods are withdrawn from the core, neutrons are not absorbed by the boron and cadmium in the control rods. This causes the chain reaction to speed up toward a supercritical reaction where the average number of neutrons fissioning ^{235}U atoms becomes higher than one per fission. When the control rods are inserted deeper into the core, a larger number of free neutrons are absorbed. This causes the chain reaction to slow down. The amount of heat generated by the reaction is regulated

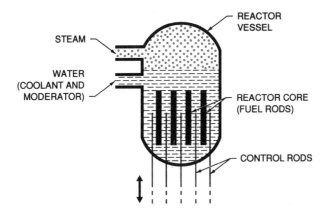

Figure 3-14. Control rods absorb neutrons and regulate the amount of fissioning in the reactor.

by the position of the control rods. Control rods are pulled out to start a reactor and dropped in to shut it down.

Reactor Vessel

A *reactor vessel* is a heavily constructed steel vessel that contains the entire reactor core. The reactor vessel houses the control rods and fuel assemblies. Coolant is pumped into the reactor vessel to remove heat from the reactor core. For maxium protection, the entire reactor core is surrounded by a reinforced concrete structure that serves as shielding. See Figure 3-15.

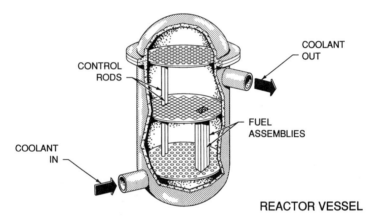

Figure 3-15. Reactor vessels are designed to withstand the heat and pressure created by nuclear fission.

Coolant

Coolant is a liquid that absorbs and transports produced heat. It is required in the nuclear reactor to remove and transport heat produced in the fissioning process. The heat in the coolant is then released as required away from the reactor core. In most reactors, water is used as a coolant. The heated water is then used for generating steam. Large nuclear reactors require as much as 330,000 gallons of water flowing through the reactor per minute to adequately remove the heat.

Shielding

Shielding is any material that contains and absorbs radiation produced in the fission process in a nuclear reactor. Shielding methods and materials vary depending on the reactor design. In all reactors, the primary purpose of the shielding is to contain the radiation and protect the environment.

NUCLEAR REACTOR TYPES

The most common type of nuclear reactor used in the United States is the light water reactor (LWR). *Light water reactors* are reactors that use a liquid coolant (water or light water) that is pumped into the reactor vessel through the reactor core to remove heat. The coolant is then pumped out of the reactor vessel and used to produce steam. The two most common LWRs are the boiling water reactor and pressurized water reactor.

Another type of reactor is the breeder reactor. A *breeder reactor* is a nuclear reactor that uses liquid sodium rather than water for heat transfer. Breeder reactors are designed to produce new fuel material as they operate.

All nuclear reactors today are fission reactors. Fusion reactors may replace fission reactors in the future. Numerous experiments have been conducted over the past 30 years without significant findings. There are no fusion reactors in operation at this time.

Boiling Water Reactors

Boiling water reactors (*BWRs*) operate with cooling water passing through the reactor core. The water is converted to steam. The steam is then used to drive steam turbo-generators for the generation of electricity. After use in the turbo-generators, the steam is returned to the core as liquid water coolant. See Figure 3-16.

Water turns to steam at 212°F. However, steam at this temperature does not have enough energy to drive a turbo-generator. To raise the heat energy in steam, the water in a BWR is kept at a pressure of 1000 pounds per square inch (psi). This prevents water from boiling and turning into steam until 545°F. At these

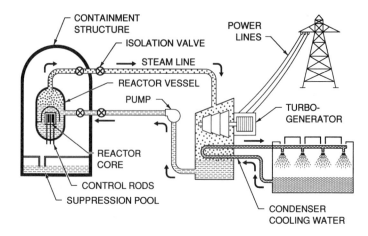

Figure 3-16. Boiling water reactors use steam created in the reactor vessel to drive the turbo-generators.

temperatures, the steam is used to drive the turbo-generators efficiently.

The total fuel mass in a large (1100 MW) BWR is approximately 187 tons. The core has a diameter of approximately 15′ and is approximately 12′ high. Control rods are inserted from the bottom instead of from the top as in a pressurized water reactor. This allows a more uniform heat generation in the vertical direction. The control rods contain four blades in the shape of a cross and operate in the spaces between fuel elements. In this configuration there is one control element for each fuel rod in the fuel assemblies.

Pressurized Water Reactors

Pressurized water reactors (*PWRs*) use a reactor core surrounded by a steel "core barrel" or "shroud" reactor vessel. The core is kept under enough pressure to prevent cooling water from flashing into steam. The reactor vessel is made of steel approximately 8″ to 9″ thick. The reactor vessel size ranges from 45′ high and 15′ in internal diameter. The reactor vessel contains the water that serves as coolant, moderator, and reflector. See Figure 3-17.

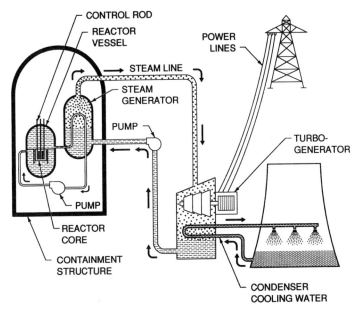

Figure 3-17. Pressurized water reactors contain a steam generator that produces steam to drive the turbo-generator.

PWRs use a double-loop system where two loops of water (primary and secondary) never mix with one another. The primary loop removes heat from the reactor core. The secondary loop provides steam to the turbo-generator.

The primary loop contains water that flows through the reactor and is pressurized to approximately 2250 psi. At this pressure, the boiling point of water is 653°F. The heated water is then pumped to the steam generator. The steam generator functions as a heat exchanger. After releasing heat to the secondary loop, water in the primary loop is returned to the reactor vessel. The water enters just above the core and flows down through the annular region (downcomer) between the shroud and the reactor vessel wall. At the bottom of the core, the water reverses direction and flows upward through the core to remove fission heat. The secondary loop water is heated as it passes through the steam generator. The heat transferred to the secondary loop in the steam generator converts the secondary loop water into steam. The steam is directed to the turbo-generator at approximately 500°F.

Some PWRs have two to four independent steam generator loops in parallel. The reactor water at 2250 psi passes through the primary loop while water at 1100 psi flows through the secondary side where it is converted into steam at about 556°F. The steam is used in a turbo-generator and then condensed back to water and returned as feedwater to the steam generator.

Breeder Reactors

Breeder reactors produce fuel material as they operate. They obtain more energy from uranium than light water reactors. In light water reactors, the fission process is determined by ^{235}U. In a breeder reactor, the chemical element plutonium is used. When struck by a neutron, the plutonium splits into two particles. Heat and neutrons are also released in the process. When these neutrons collide with other plutonium atoms, the fission process is sustained. See Figure 3-18.

In a liquid metal fast breeder reactor (LMFBR), liquid sodium is used as a coolant. Liquid sodium allows the reactor to operate at very high temperatures and low pressures. The LMFBR uses two loops that function similarly to a pressurized water reactor. Liquid sodium is pumped through the reactor core into the heat exchanger. Heat from the core is transferred to another loop similar to a pressurized water reactor. The liquid sodium transports heat in the secondary loop to a steam generator. Steam is then directed to the turbines, which drive the generators.

Plutonium used in a breeder reactor does not occur naturally. It is a product of uranium used as a fuel. When ^{238}U is struck by a neutron, the atom absorbs the neutron and forms an atom of plutonium. This occurs as a by-product in a light water reactor. Breeder reactors are designed to enhance this process. Commercial breeder reactors are in use in France, Great Britain, the Soviet Union, and the Federal Republic of Germany (West Germany). There are no commercial breeder reactors in operation in the United States at this time.

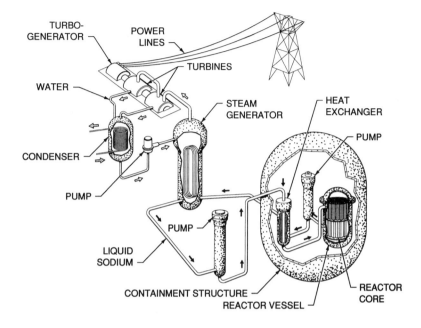

Figure 3-18. Breeder reactors contain two separate loops to transport heat from the reactor vessel to the turbo-generator.

Radiation

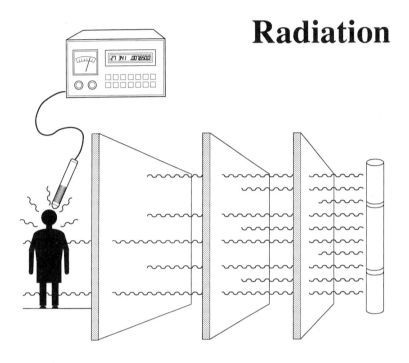

Radiation is the emission of high-energy particles (rays) from the unstable nuclei of atoms. The three common types of radiation in nuclear power plants are alpha, beta, and gamma rays. Each of these have different weights, electrical charges, penetrating ability, and speed.

Radiation is measured in roentgens, rads, and rems. The degree of exposure to radiation is determined by the distance from the source, length of exposure time, shielding, and source intensity. A curie is the unit used to measure the rate at which particles or rays are released from radioactive isotopes. Metric prefixes are used with standard measurements throughout the nuclear power industry.

RADIATION SOURCES

Radiation occurs in nature from earth, rocks, and cosmic rays from outer space. Radiation also occurs from sources related to manufacturing and technology such as X rays, radio waves, microwaves, and cathode-ray tubes in televisions and computers. Other such sources include coal combustion, luminous watches, and nuclear energy. These amounts of radiation are distributed over a wide area and a great period of time.

The average dose of radiation for a United States citizen is estimated to be 160.81 mrem (millirem or thousandths of a rem) per year. The total radiation exposure received in the United States is approximately 53% (85 mrem) from natural sources and approximately 47% (75.81 mrem) from manufactured sources. Of the manufactured sources of radiation, only 0.01 mrem comes from the nuclear power industry. See Figure 4-1.

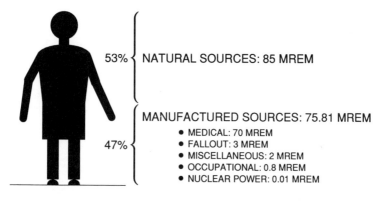

Figure 4-1. The average dose of radiation for a United States citizen is approximately 160.81 mrem per year.

RADIATION TYPES

Radioactivity is the process of spontaneous disintegration of radioactive nuclei. *Radiation* is the emission of high-energy particles (rays) from atoms with unstable nuclei. The four types of radiation are alpha radiation, beta radiation, gamma radiation, and neutron radiation. Symbols used for alpha, beta, and gamma radiation are the first three letters of the Greek alphabet: alpha (α), beta (β), and gamma (γ). The symbol for neutron radiation is n. All types of radiation are potentially harmful and differ in their content, weight, electrical charge, penetrating ability, speed, and the type of hazard they present. See Figure 4-2.

ALPHA, BETA, AND GAMMA RADIATION			
Symbol	α	β	γ
Content	Particles. Nuclei of helium atom	Electrons emitted when neutrons in nuclei change to protons	Electromagnetic waves or photons
Weight (mass)	Heavy. Four times hydrogen atom	Light. $\frac{1}{1840}$ as heavy as hydrogen atom	None
Electrical charge	Positive (+2)	Negative (–1)	No charge
Penetrating ability	Stopped by several sheets of paper	Stopped by 1/8" thick aluminum	Stopped by 24" water, 12" concrete, 4" lead
Speed	2000 to 20,000 miles/sec	46,000 to 170,000 miles/sec	186,000 miles/sec (speed of light)
Type of hazard	High internal hazard	Moderate internal or external hazard	High external hazard

Figure 4-2. Alpha (α), beta (β), and gamma (γ) radiation are the three most common types of radiation in nuclear power plants.

Radiation is the result of energy emitted from an unstable nucleus of an atom as it regains stability. For example, an unstable atom can be compared to a rock balanced on top of a hill. The rock has potential energy. If the rock is allowed to roll down the hill, the potential energy will be released. When the rock reaches the bottom of the hill, all of the potential energy has been released and the rock is in a stable state. This example is similar to an atom with a nucleus that changes from an unstable to a stable state. The potential energy present in the nucleus while in the unstable state is released as it changes to a stable state.

Radioactive decay is loss of radioactivity by radioisotopes over a period of time and is measured in half-lives. During this process, the nucleus changes in stability into the form of another isotope or chemical element. Uranium and other naturally radioactive chemical elements decay spontaneously and add to background radiation. *Background radiation* is radiation present in the Earth's atmosphere.

Alpha, beta, and gamma rays travel at different speeds. Their ability to penetrate materials also varies. Alpha rays are the slowest and have the least penetrating ability. Beta rays are heavier and travel faster than alpha rays. Gamma rays have more penetrating ability than alpha or beta rays. See Figure 4-3.

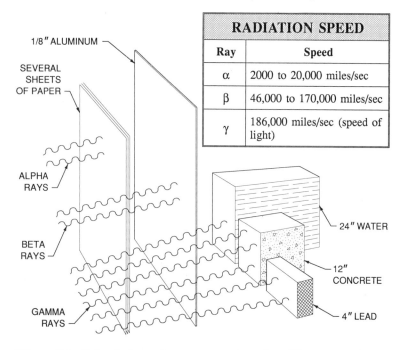

RADIATION SPEED	
Ray	**Speed**
α	2000 to 20,000 miles/sec
β	46,000 to 170,000 miles/sec
γ	186,000 miles/sec (speed of light)

Figure 4-3. Alpha, beta, and gamma rays travel at different speeds and have different penetrating abilities.

Alpha Radiation

Alpha radiation is the most energetic (densely ionizing) but the least penetrating type of radiation. Alpha radiation is in the form of alpha particles. The alpha particles can be stopped by a sheet of paper. An alpha particle is the same as a helium atom without the two electrons. Alpha particles are the largest of the particles emitted by radioactive isotopes.

Alpha particles consist of two protons and two neutrons that act as one particle. For every alpha particle emitted from the nucleus of a radioactive atom, two protons and two neutrons are emitted. An *alpha particle* is a positive-charged nuclear particle identical to a helium atom nucleus, which has two protons and two neutrons. See Figure 4-4.

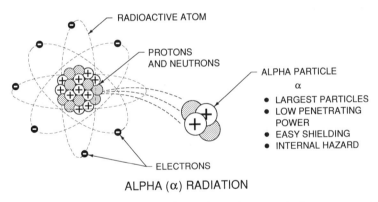

ALPHA (α) RADIATION

Figure 4-4. Alpha particles are large, have low penetrating power, and are easy to shield against.

Atoms emitting alpha particles change to an atom of a different chemical element because the atomic number is reduced by two. For example, ^{238}U has an atomic number of 92. There are 92 protons and 146 neutrons in ^{238}U. It has a mass number of 238 (92 protons + 146 neutrons = 238). This is also expressed as $^{238}_{92}U$. When a uranium nucleus emits an alpha particle, it loses two protons and two neutrons. After it loses an alpha particle, the nucleus has 90 protons and 144 neutrons. With an atomic number of 90, the atom is no longer uranium but thorium. This results in the formation of the isotope thorium 234 (^{234}Th), which has a mass number of 234 (90 protons + 144 neutrons = 234) and an atomic number of 90. Because each proton has a positive electrical charge, the total charge of alpha particles is +2.

Alpha particles are the heaviest and slowest moving of all radiation particles. Alpha particles can travel a few inches in the air, but cannot penetrate the skin. However, alpha particles emitted from radioactive material can enter the body through open wounds, food ingested, or air breathed.

Beta Radiation

Beta radiation consists of high-speed electrons emitted from the unstable nuclei of certain radioactive chemical elements. See Figure 4-5. Beta particles are sometimes difficult to comprehend because electrons have a negative electrical charge. However, after scientists discovered that protons and neutrons can disintegrate into other particles, positrons were discovered. A *positron* is a particle that has the same mass as an electron but contains a positive electrical charge. In addition, it was also discovered that a proton can be transformed into a positron and a neutron.

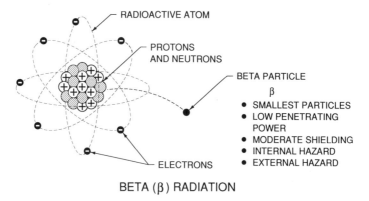

BETA (β) RADIATION

Figure 4-5. Beta particles are small, have low penetrating power, and are easy to shield against.

The positron exists until colliding with an electron. A neutron can be split into a proton and an electron. The proton remains in the nucleus. This raises the atomic number by one, but the mass number remains the same. The electron is emitted from the nucleus as a beta ray. This process can be reversed. For example, a proton and an electron can join to form a neutron. A neutron and a positron can join to form a proton.

Beta particles have a negative electrical charge and are approximately $\frac{1}{1840}$ times as heavy as a hydrogen atom. Beta particles have positrons. A positron contains the same positive electrical charge as a proton and the same small mass as an electron. A nucleus that emits a positron also emits a neutrino. A *neutrino* is an uncharged (neutral) particle with high penetrating power that is produced in a nuclear reaction. A nucleus that emits a beta particle with a negative electrical charge also emits an antineutrino. An *antineutrino* is an antiparticle that has no mass or electrical charge. The electron and the antineutrino are emitted from the nucleus as they form. However, the proton remains in the nucleus. The resulting nucleus has one more proton and one less neutron. See Figure 4-6.

Chemical elements can be changed by beta radiation. For example, ^{238}U can be changed to plutonium 239 (^{239}Pu) for use in a breeder reactor. Two electrons are emitted during this

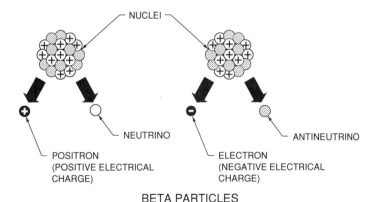

BETA PARTICLES

Figure 4-6. Beta particles have either positive or negative electrical charges.

process. A neutron collides with a ^{238}U nucleus and is absorbed. This results in the isotope uranium 239 (^{239}U), which is radioactive and emits a beta particle. The beta particle is formed as one of the neutrons in the nucleus changes to a proton and a beta particle. The nucleus now has one more proton and one less neutron. Its atomic number is now 93 and it is now neptunium 239 (^{239}Np). The neptunium isotope is radioactive and emits another beta particle. This reduces the number of neutrons by one again, which raises the atomic number to 94. Plutonium 239 has an atomic number of 94 ($^{238}_{92}$U + 1 neutron and – 2 electrons → $^{239}_{94}$Pu).

A series of changes are required before ^{238}U reaches a stable state. For example, ^{238}U changes to isotopes of thorium, radium, radon, polonium, and lead. All of these isotopes are radioactive except lead 206 (^{206}Pb).

Beta radiation is a more penetrating type of ionizing radiation than alpha radiation, but has a lower ionization ability. Like alpha radiation, it has its most serious effects when inhaled or ingested. Beta radiation is more common in nuclear reactors than alpha radiation but less frequent than gamma radiation. Beta radiation is more easily shielded against than gamma radiation. Aluminum at least ⅛″ thick is sufficient to shield beta radiation.

Gamma Radiation

Gamma radiation consists of high energy electromagnetic energy waves. It has more penetrating power than alpha radiation or beta radiation. Gamma radiation commonly occurs with beta radiation emission. Gamma rays are similar to X rays but have higher energy and consequently can cause more biological damage to human tissue. Gamma rays are released from a nucleus in a high-energy state after radioactive decay. Approximately 95% of the radiation produced in a nuclear facility are gamma rays.

Gamma rays are a type of photon. A *photon* is a particle of electromagnetic radiation. It travels at the speed of light, which is 186,000 miles per second. Electromagnetic radiation occurs when some atoms or nuclei that are in an unstable state change to a stable state without changing atomic weight or mass number. In this process, a photon is emitted. The primary difference between photons and alpha and beta particles is the speed at which they travel. Electrons, protons, and other particles have mass and travel at various speeds but never travel at the speed of light. See Figure 4-7.

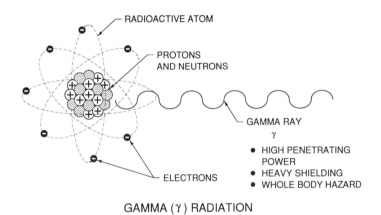

GAMMA (γ) RADIATION

Figure 4-7. Gamma rays are pure energy, have no electrical charge, and are highly penetrating.

Photons have no mass and are created when an unstable nucleus or atom changes to a more stable state. They travel at the speed of light until they are absorbed by other atoms or nuclei. This results in the atoms or nuclei becoming less stable than before. Photons can also be absorbed by an alpha or a beta particle, causing the particle to go faster.

Atoms emit photons when the outer electrons change their orbits or speeds, which puts the atom in a more stable state. In some instances, an electron that moves from an orbit farther from the nucleus to a closer orbit also puts the atom into a more stable state. The atom then releases excess energy in the form of a photon. The photon may be light, heat, or other types of rays depending on the electron involved and the orbital position.

Nuclei emit photons when the nucleus, not the atom as a whole, is unstable. For example, when a radium atom decays, it emits an alpha particle. This changes the nucleus to the gas radon. The alpha particle can travel at a fast or slow speed. If the particle is traveling at a fast speed, all excess energy of the radium has been used to propel the particle. If the particle is traveling at a slow speed, the radon nucleus is left with some excess energy that causes the nucleus to be unstable. The nucleus can regain stability by releasing the excess energy in the form of a gamma ray photon. See Figure 4-8.

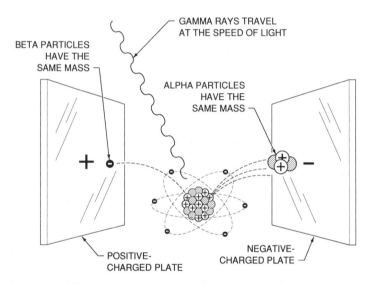

Figure 4-8. All gamma rays travel at the speed of light.

Photons behave like particles or act in waves, depending on the reaction. Light, heat, and gamma rays are all composed of photons. These photons all move at the same speed, but when acting as a particle some photons carry more energy than others. When acting in waves, they occur at different wavelengths. Photons having the highest energy have the shortest wavelengths. See Figure 4-9.

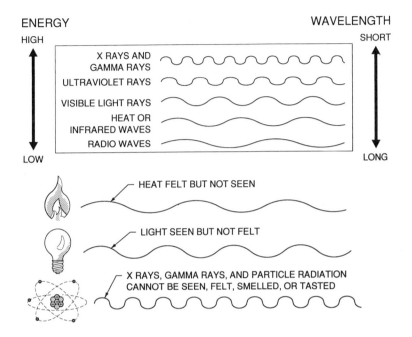

Figure 4-9. Photons with high energy have shorter wavelengths.

Neutron Radiation

Neutron radiation is the most highly penetrating and difficult to shield of all the radiation types. It is produced by the fission process. Neutrons are contained within the reactor vessel and are rarely encountered in the nuclear facility. The most common source of neutrons in the nuclear facility is from the neutron source used for instrument testing and reactor startup. See Figure 4-10.

HALF-LIFE

Half-life is a unit of time used by scientists to measure radioactive decay. *Half-life* is the time required for a quantity of radioactive material to lose one half its radioactivity. During this process, half of the atoms decay into another chemical element or isotope.

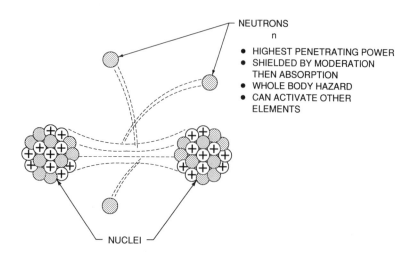

NEUTRON (n) RADIATION

Figure 4-10. Neutron radiation is produced by the fission process.

For example, radioactive thorium has a half-life of 8×10^4 years ($8 \times 10 \times 10 \times 10 \times 10 = 80,000$ years). Half of the atoms in one pound of thorium will decay into other chemical elements in 80,000 years and half of the thorium will remain. Half-life is extremely important when considering safety precautions and storage of nuclear waste material.

Different radioisotopes have different half-lives. Half-lives of radioisotopes can vary from fractions of a second to billions of years depending on their chemical makeup. For example, ^{211}Pb has a half-life of 36.1 minutes. See Figure 4-11.

Most natural-forming radioisotopes have very long half-lives. Scientists theorize that all radioisotopes were present on Earth at one time. Over the years, those radioisotopes with short half-lives have decayed. Those radioisotopes with very long half-lives have remained. Radioisotopes that have short half-lives are produced artificially by bombarding nuclei with fast-moving nuclear particles such as neutrons in nuclear reactors.

HALF-LIFE

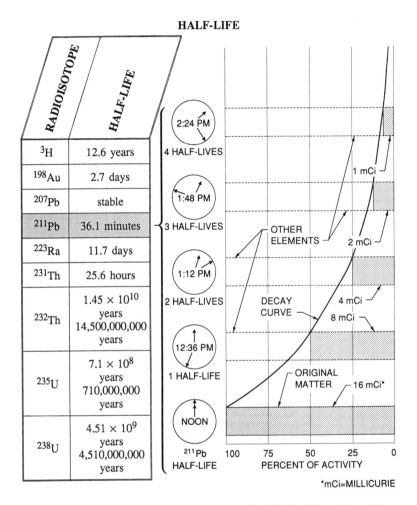

Figure 4-11. Half-life is the time required for half of the atoms in a radioisotope to decay.

RADIATION MEASUREMENT

For maximum safety when working with radioactive materials standard measurements are used. Standard measurements are used throughout the nuclear power industry for specifying proper radiation limits.

Each unit of standard measurements uses metric prefixes to express the amount measured. Metric prefixes allow expression of measurement units using powers of 10. For example, a milliroentgen is equal to $\frac{1}{1000}$ of a roentgen. *Milli* is the metric prefix for one thousandth. Using metric prefixes reduces the number of digits required to express a measurement. Radioactivity is measured in curies. Radiation is measured in roentgens, rads, and rems. See Figure 4-12.

METRIC PREFIXES				
Prefix	**Symbol**	**Power of 10**	**Unit**	**Example**
pico	p	10^{-12}	$\frac{1}{1,000,000,000,000}$	1 picofarad = 10^{-12} farad
nano	n	10^{-9}	$\frac{1}{1,000,000,000}$	1 nanosecond = 10^{-9} second
micro	μ	10^{-6}	$\frac{1}{1,000,000}$	1 microcurie = 10^{-6} curie
milli	m	10^{-3}	$\frac{1}{1000}$	1 millirem = 10^{-3} rem
centi	c	10^{-2}	$\frac{1}{100}$	1 centimeter = 10^{-2} meter
deci	d	10^{-1}	$\frac{1}{10}$	1 decigram = 10^{-1} gram
kilo	k	10^{3}	1000	1 kilowatt = 10^{3} watts
mega	M	10^{6}	1,000,000	1 megaohm = 10^{6} ohms
giga	G	10^{9}	1,000,000,000	1 gigaelectronvolt = 10^{9} electronvolts

Figure 4-12. Metric prefixes are used with standard measurements. Curies measure radioactivity (emission of radiation). Roentgens, rads, and rems measure radiation (absorption of radiation).

Curies

A *curie* is the unit used to measure the rate at which particles or rays are released from a radioactive isotope. A curie (Ci) is the quantity of radioactive material in which 3.7×10^{10} atoms are decaying every second. This amount is based on the radioactivity of radium as a standard. In 1 gram of radium, 3.7×10^{10} atoms decay every second. This means 1 gram of radium contains 1 curie of radioactivity. The equivalent amount of 1 curie of another chemical element other than radium may require more or less than 1 gram. For example, 1 curie of phosphorus 32 (^{32}P) weighs 20 grams. The number of curies is determined by how fast unstable atoms emit particles or photons.

When using radiotracers, four atoms decaying per second can be easily counted. However, the smaller units millicurie (mCi) and microcurie (µCi) are used more frequently. A millicurie is equal to $\frac{1}{1000}$ of a curie. A microcurie is equal to $\frac{1}{1,000,000}$ of a curie. Curies were originally used for measuring emissions of gamma rays only. Other more accurate methods have been developed since that time.

Roentgens

A roentgen measures X rays and gamma rays. A *roentgen* (pronounced *rent′ ghen*) is the quantity of X rays or gamma radiation that will produce one electrostatic unit of positive or negative electrical charge in a cubic centimeter of dry air. The abbreviation for roentgen is r. The roentgen is the measure of the gamma radiation present required to produce a standard amount of ionization of the air. Ionization of the air occurs when photons knock electrons out of atoms of nitrogen, oxygen, and other chemical elements present in the air. Ionization of the air results in negative- and positive-charged atoms or molecules, and other positive- and negative-charged particles.

Three limitations to consider when defining a roentgen include the following: (1) ionization caused by particle radiation produced by neutrons and alpha particles, (2) ionization produced in substances other than air such as hydrogen, lead, and human tissue, and (3) although ionization is produced in a cubic centimeter of air, for high-energy photons the electrons produced by the photons in a certain volume of air do not stay in that volume. These electrons themselves have a lot of energy. The roentgen is only a measure of the ionization produced by the radiation. It does not measure how much energy was required to produce the ionization or whether the ionized particles will stay in the given area.

Rads

A *rad* (radiation absorbed dose) is a unit of absorbed dose equal to 100 ergs of energy per gram. It is a measurement of the ionizing radiation absorbed per gram of material. When radiation passes through a material, energy is removed from the radiation. The energy absorbed by the material can result in molecules split into atoms, ionization of atoms, and other reactions. The relationship between a rad and a roentgen is 1 roentgen in air equals 0.88 rad in water.

Rems

A *rem* (roentgen equivalent in man) is a radiation dose equivalent in biological damage to 1 rad of 250 kW of X rays. It is a measurement of the amount of biological damage from a dose of radiation. A rem quantifies the biological damage produced by the absorbed dosage of 1 rad into body tissue. A rem takes into account both the amount of radiation deposited in body tissues and the type of radiation—alpha, beta, or gamma radiation.

Biological damage from radiation depends on the energy transmitted to the cell and its constituents and on the number of cells struck. These in turn depend on the type of radiation and the dose that makes up the total amount of radiation energy absorbed by the affected tissue.

Different radiation sources are factored using quality factors when determining rem. *Quality factors* are standards determined by the amount of biological damage a specific radiation can produce. The formula for finding rem is:

Rem = rad × QF
Quality Factors (QF)
Beta = 1
Gamma = 1
Thermal neutrons = 1
Fast-moving neutrons = 10
Alpha radiation = 20

In this formula, 1 rem is equal to the rad amount times the quality factors. For example, how many rem are produced by 300 rad of gamma radiation?

Rem = rad × QF
Rem = 300 × 1
Rem = 300

RADIATION EXPOSURE

The level of exposure to radiation is determined by the distance from the source, the length of exposure, shielding, and source intensity. Radiation exposure increases closer to the radiation source. Moving away from the radiation source, the radiation exposure decreases. The amount of radiation exposure received uses the mathematical concept of inverse square. The intensity of radiation exposure is proportional to the inverse square of distance from the radiation source. This principle is stated mathematically as

$$Rems/hour = \# \ of \ rems/hour \ at \ 1' \times \frac{1}{distance^2}$$

For example, a person is standing 1′ from a radioisotope and receives radiation at a rate of 100 rems/hour. If the person moves to 2′ away from the radioisotope, how many rems/hour would be received?

$$Rems/hour = \text{\# of rems/hour at } 1' \times \frac{1}{distance^2}$$

$$Rems/hour = 100 \times \frac{1}{2^2}$$

$$Rems/hour = 100 \times \frac{1}{4}$$

Rems/hour = 25

The amount of radiation exposure decreases proportionally as a person moves away from the radiation source. See Figure 4-13.

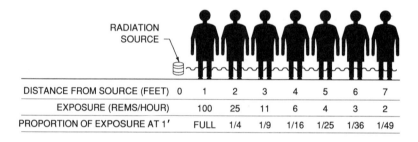

DISTANCE FROM SOURCE (FEET)	0	1	2	3	4	5	6	7
EXPOSURE (REMS/HOUR)		100	25	11	6	4	3	2
PROPORTION OF EXPOSURE AT 1'		FULL	1/4	1/9	1/16	1/25	1/36	1/49

$$REMS/HOUR = \text{\# OF REMS/HOUR AT } 1' \times \frac{1}{DISTANCE^2}$$

Figure 4-13. The amount of radiation exposure decreases proportionally with the distance.

The principle of inverse square works only when the radiation is concentrated in a specific location or point source. A *point source* is a concentration of radioactivity in a small volume. The principle of inverse square will not work in cases of large areas of radiation.

Shielding between the radiation source and the person will reduce the amount of radiation exposure. In most cases, elements of materials with a high atomic number function as the best shielding material. Lead is an example of a good shielding material as it has an atomic number of 82.

Alpha and beta particles and gamma rays can move through a shielding material without colliding with an electron. Shielding material used is not designed to stop all penetration, but reduce the level of radiation that can be hazardous to personnel. Shielding is rated by the amount of material required to stop half of the radiation from penetrating through the shielding material. A *half-value layer* is the thickness of shielding material that will stop half of the radiation from penetrating through. See Figure 4-14.

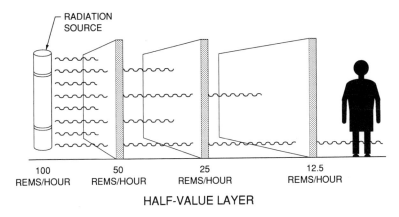

Figure 4-14. A half-value layer of material will stop 50% of the radiation.

Chapter 5

Nuclear Power Plant Safety

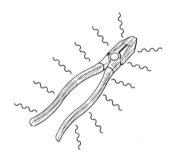

DANGER

HIGH RADIATION AREA

All nuclear power plants operating in the United States must comply with regulations stipulated by various agencies of the United States government. The Atomic Energy Act of 1954 was established to develop nuclear power and protect the public. All current laws and regulations governing nuclear power plants today have their beginning in this act.

Construction permits and operating licenses must be granted by the Nuclear Regulatory Commission (NRC) before construction of a nuclear power plant can begin. Quality assurance criteria is required by the NRC to ensure safety-related procedures. In addition to setting upper limits of radiation doses, the NRC requires licensees to minimize radiation exposure of workers in nuclear power plants. A wide variety of radiation detection and measurement devices are used so that workers may be protected.

REGULATIONS

Of the approximately 240 nuclear power plants operating in the world today, the United States is the leader with 110 plants. Nuclear power plants use less fuel than fossil fuel power plants. One short ton of uranium produces as much heat energy as the burning of three million short tons of coal or 12 million barrels of oil. A *short ton* is 2000 pounds and is equivalent to 0.907 185 metric tons (t).

Nuclear power plants emit very little pollution into the air during use. However, nuclear power plants cost more to build than fossil fuel power plants. Because nuclear power plants are potentially dangerous, the construction and operation requirements are much more stringent than those of fossil fuel power plants.

All nuclear power plants built in the United States must comply with regulations stipulated by various agencies of the United States government. Failure to comply with these regulations can result in construction delay or cancellation of the project or plant licensing.

Atomic Energy Act of 1954

The *Atomic Energy Act of 1954* was established by the federal government to develop nuclear power and protect the health and safety of the public by ensuring a minimum impact on the environment. The U.S. Atomic Energy Commission (AEC), which was formed in 1946, administered the act. With the passage of the Energy Reorganization Act in 1974, the responsibilities of the AEC were divided into the Nuclear Regulatory Commission (NRC) and the Energy Research and Development Administration (ERDA). All current laws and regulations governing nuclear power plants today have their beginnings in the Atomic Energy Act of 1954. See Figure 5-1.

The Code of Federal Regulations (CFR) is written in the Atomic Energy Act of 1954. Laws and regulations pertaining to nuclear power plants are listed in the CFR under Title 10, "Energy," Chapter 1 - Regulations. In Chapter 1, specific parts

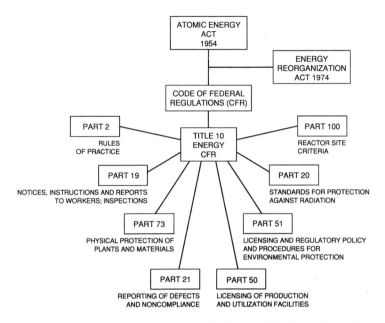

Figure 5-1. All current laws and regulations regarding nuclear power plants began with the Atomic Energy Act of 1954.

detail regulations pertaining to nuclear power plant design, construction, and operating procedures. For example, Part 19 of CFR 10, Chapter 1 (10CFR Part 19) covers Notices, Instructions and Reports to Workers; Inspections. Parts of 10CFR that are relevant to nuclear power plant licensing, design, construction, and operating procedures include Parts 2, 19, 20, 21, 50, 51, 73, and 100.

10CFR Part 2 - Rules of Practice pertains to the issuing, suspending, revoking, and other actions regarding licensing. Construction permits are also covered in Part 2.

10CFR Part 19 - Notices, Instructions and Reports to Workers; Inspections pertains to working conditions for persons who work with radioactive material that requires a license from the Nuclear Regulatory Commission.

10CFR Part 20 - Standards for Protection Against Radiation pertains to requirements that ensure protection from radioactive materials. This part also includes information regarding personnel radiation doses and control of radioactive material releases to "as low as reasonably achievable" (ALARA).

10CFR Part 21 - Reporting of Defects and Noncompliance pertains to the reporting procedures of noncompliance with established design, construction, and operating procedures specified by federal regulations. This part also provides the procedures that assure that discrimination cannot take place as a result of a noncompliance report being filed. See Figure 5-2.

10CFR Part 50 - Licensing of Production and Utilization Facilities pertains to specific licensing, design, and construction requirements of nuclear power plants. This part also covers plant quality assurance criteria, emergency plan requirements, and the recordkeeping of safety-related items.

10CFR Part 51 - Licensing and Regulatory Policy and Procedures for Environmental Protection pertains to the environmental impact of nuclear power plants and the procedures for public notices and hearings regarding the environment.

10CFR Part 73 - Physical Protection of Plants and Materials pertains to the security of nuclear power plants for protection of nuclear materials against acts of industrial sabotage.

10CFR Part 100 - Reactor Site Criteria pertains to the selection of the site for nuclear power plants.

Nuclear Regulatory Commission

The *Nuclear Regulatory Commission (NRC)*, established in 1974 as an independent agency of the federal government, enforces regulations to protect the public health and safety. Responsibilities of the NRC that pertain to the nuclear power industry include development and enforcement of laws regarding plant design, and the manufacture, distribution, and storage of radioactive materials. The NRC also is involved with the protection of the environment as it relates to nuclear power plants.

Department of Energy

The *Department of Energy (DOE)*, established in 1977, has responsibilities acquired from the Atomic Energy Commission

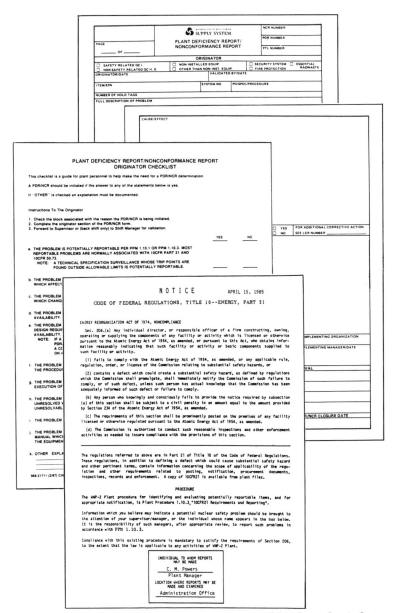

Washington Public Power Supply System

Figure 5-2. The Code of Federal Regulations (CFR) includes procedures for documenting noncompliance. Local nuclear power plants develop checklists for noncompliance reports.

(1946 to 1974) and the Energy Research and Development Commission (1974 to 1977). Responsibilities of the DOE that pertain to the nuclear power industry include waste management and research and development in nuclear science. The primary function of the DOE in the nuclear power industry is the management of spent nuclear fuel, radioactive waste, and uranium mill tails.

Environmental Protection Agency

The *Environmental Protection Agency* (*EPA*), established in 1970, has responsibilities in the nuclear power industry that include initiating federal guidelines for creating and establishing standards affecting public health and the environment. This includes development of the criteria for handling and disposing of all radioactive wastes, including short-term and long-term storage.

Department of Transportation

The *Department of Transportation* (*DOT*), as it pertains to the nuclear power industry, regulates the shipment of all privately owned radioactive materials and radioactive waste. This includes the identification and classification of all radioactive material shipped by any mode of transportation.

American National Standards Institute

The *American National Standards Institute* (*ANSI*) develops standards for design, construction, and procedures as they relate to nuclear power plants. ANSI employs professionals in the nuclear power industry who develop standards in conjunction with the consensus of experts in their respective fields. ANSI standards are commonly used as guidelines and are required if they are stipulated for a specific contract or purchase order.

ANSI 18.7, *Administrative Controls and Quality Assurance for the Operational Phase of Nuclear Power Plants,* includes information regarding the requirements and recommendations for activities such as inspection, testing, repairing, and maintenance procedures.

NUCLEAR POWER PLANT LICENSING

Nuclear power plant licensing is administrated nationally under the control of the Office of Nuclear Reactor Regulation. Before a nuclear power plant can be built, a construction permit and an operating license must be obtained from the NRC. This is a lengthy process requiring licenses and permits from the federal, state, and local governments. These licenses and permits determine the feasibility and impact of the proposed nuclear power plant. To obtain licenses and permits, the following criteria must be substantiated:

- The proposed nuclear power plant is in compliance with the National Environmental Policy Act.
- The utility company is qualified to design, construct, and operate the proposed nuclear power plant.
- Construction and operation of the nuclear power plant will not cause undue risk to the public health and safety.
- Licensing of the nuclear power plant will not be harmful to the national defense or security.

Throughout the licensing process, there is a required procedure that provides for involving the public. This provides an opportunity for every member of the public to voice an opinion.

Since the first operation of a commercial nuclear power plant, the United States has accrued in excess of 1000 reactor years of operating experience. This experience has occurred without a single death to a member of the public caused by radiation. Nuclear power plant licensing is covered under *10CFR Part 50 - Licensing of Production and Utilization Facilities*. In addition, *10CFR Part 51 - Licensing and Regulatory Policy and Procedures for Environmental Protection* was added to meet the requirements of the National Environmental Policy Act of 1969. This act specifies the procedures for processing environmental impact statements.

RADIOACTIVE MATERIAL

In nuclear power plants, radioactive material can exist in solid, liquid, and gaseous states. Solid radioactive material includes materials such as fuel and activated or contaminated materials. Liquid radioactive materials include coolant used in the reactor, chemical waste, and contaminated laundry waste. Gaseous radioactive material includes activated oxygen, nitrogen, and argon. Also, there are gases that are the result of the fission process such as krypton, iodine, and argon.

Contamination is the presence of removable radioactive material in any place where it is not desired. *Decontamination* is the removal of radioactive material from an undesired place. Nuclear power plants have a high potential for contamination of materials. It is always the best policy to assume spills and leaks are contaminated and notify the proper personnel. If a metal tool were exposed to an intense beam of neutrons, some of the atoms in the tool would become radioactive. If a tool were left on the floor and splashed with radioactive process water, it would become contaminated. The difference between an object being radioactive or contaminated is that a radioactive object is radioactive from the inside. No amount of wiping or washing will make it clean or nonradioactive. Contamination is removable radioactivity in which the contaminated object can be cleaned using a solvent or soap in a designated area. See Figure 5-3.

Designated areas of a nuclear power plant are identified as potentially contaminated areas. These areas require a radiation work permit and protective clothing for entry. No tools or other material in these areas should be touched with the bare hands or removed from the area without following the proper procedure. Carelessness with contaminated materials can spread contamination throughout the plant. For example, contamination in the lunchroom could result in the potential for ingesting or inhaling the contaminated material. The actual radiation from the contaminated material may be quite low. However, even a very small amount of contaminated material, if taken into the body, could be a serious problem. Contamination must be treated with a great deal of respect.

Warning: Follow plant procedures when working with contaminated materials.

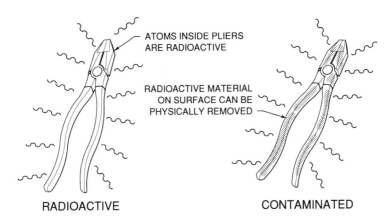

ATOMS INSIDE PLIERS
ARE RADIOACTIVE

RADIOACTIVE MATERIAL
ON SURFACE CAN BE
PHYSICALLY REMOVED

RADIOACTIVE   CONTAMINATED

Figure 5-3. Radioactive objects are radioactive from the inside. Contaminated objects have radioactive material on the surface.

PROTECTION FROM RADIATION

Nuclear power plants are required by federal regulations to maintain a safe environment for workers. Current standards recommend that workers be exposed to no more than 5000 mrem above background level in a given year. The recommended limit for the general public is 25 mrem above the background level in a given year. Workers in nuclear plants wear radiation measurement instruments that monitor the exposure to radiation. These devices are checked regularly to record exposure to radiation. Special equipment, clothing, and limiting exposure to radioactive material all serve to minimize exposure to radiation.

The NRC stipulates policies and procedures that minimize the potential hazards of working in a nuclear power plant. Health physics personnel is responsible for checking for potential radiation hazards. In addition, all contamination cases are reported to health physics.

Quality Assurance Criteria

Quality assurance criteria is required by the NRC to ensure safety-related procedures in nuclear power plants. The NRC was created to regulate the nuclear power industry to ensure that the health and safety of the public are protected. *10CFR Part 50 - Licensing of Production and Utilization Facilities, Appendix B* details the quality assurance criteria for nuclear plants. This information covers the safety-related procedures required in nuclear power plants. See Figure 5-4.

ALARA

The NRC, in addition to setting upper limits for radiation doses, also requires licensees to minimize occupational exposures to workers at nuclear power plants below the recommended dose limits. According to *10CFR Part 20 - Standards for Protection Against Radiation*, to provide maximum protection from the potential hazards of radiation, "conduct all operations involving radiation and radioactive material such that radiation exposure to employees, contractors, and the general public is maintained *as low as reasonably achievable (ALARA)* below the limits specified in *10CFR Part 20*." The policy of ALARA requires the effort of all workers in nuclear power plants. Exposure to radiation can be reduced using the following methods:

- Reduce the strength of the radiation source.
- Increase the amount of shielding.
- Increase the distance between workers and radioactive material.
- Reduce the exposure time to the radioactive material.

QUALITY ASSURANCE CRITERIA

I. Organization
- Assign responsibility
- Establish authority
- Assure independence

II. Quality Assurance Program
- Document program
- Identify items covered
- Develop procedures
- Provide management review

III. Design Control
- Document all design requirements
- Review material and processes
- Identify and control interfaces
- Control design changes

IV. Procurement Document Control
- Include all applicable requirements
- Invoke NRC quality requirements as applicable

V. Instructions, Procedures and Drawings
- Document quality activities
- Include acceptance criteria

VI. Document Control
- Assure use of proper documents
- Review and approve documents
- Control document changes

VII. Control of Purchased Material, Equipment and Services
- Assure conformance to procurement documents
- Evaluate suppliers
- Provide for inspection and audit as applicable
- Provide documentary evidence of conformance

VIII. Identification and Control of Materials, Parts and Components
- Provide item identification and control
- Assure prevention of use of incorrect or defective items

IX. Control of Special Processes
- Assure welding, heat treating and NDE are controlled
- Provide qualified personnel and procedures
- Use applicable codes and standards

X. Inspection
- Establish, document and execute inspection programs
- Provide independent inspectors
- Provide process monitoring when appropriate
- Recognize mandatory hold points

XI. Test Control
- Establish, document and execute test programs
- Include test requirements and acceptance criteria
- Document and evaluate test results

XII. Control of Measuring and Test Equipment
- Provide periodic calibration, adjustment and control

XIII. Handling, Storage and Shipping
- Control handling, storage, shipping, cleaning and preservation
- Use special protective environments when necessary

XIV. Inspection, Test and Operating Status
- Status labels, tags and cards for:
  - Inspections and tests
  - Operating status of components and systems

XV. Nonconforming Materials, Parts or Components
- Control nonconforming materials, parts or components
- Review and disposition in accordance with documented procedures

XVI. Corrective Action
- Promptly identify and correct adverse conditions
- Document and report corrective actions

XVII. Quality Assurance Records
- Prepare, maintain and retain QA records
- Provide for records identification and retrievability

XVIII. Audits
- Provide planned and periodic audits
- Perform audits-by trained and independent personnel
- Management review of documented audit results
- Follow-up as necessary

Figure 5-4. The NRC's quality assurance criteria covers safety-related procedures in nuclear power plants.

Radiation Measurement Instruments

Radiation cannot be detected by human senses. Radiation measurement instruments are used to indicate the presence (detect) and measure the amount of radiation in nuclear power plants so that precautions may be taken to protect workers. A *dosimeter* is any device that measures doses of radiation. All persons in controlled areas of nuclear power plants are required to wear a dosimetry device to measure radiation. *Dosimetry* is all equipment used to detect and measure radiation emitted and doses received. Radiation measurement instruments constantly monitor the amount of radiation present in different areas of the nuclear power plant.

Geiger Counter. The Geiger-Muller (G-M) or Geiger counter is the most common and most frequently used of all radiation measurement instruments (dosimeters). A *Geiger counter* is an instrument that detects the presence and measures the intensity of radiation through the ionizing effect of an enclosed gas. The main component of the Geiger counter is a glass tube filled with gas at low pressure and connected to an electric meter. When alpha or beta particles or gamma ray photons enter the gas, an electric current is produced. The amount of current produced is registered as a visible or audible signal. Visible signals are registered on the meter and audible signals increase in intensity as the radiation level increases. Geiger counters always register a small amount of radiation. This is background radiation and indicates the Geiger counter is operating. Background radiation readings between 0 and 0.1 mrem/hour are acceptable. See Figure 5-5.

Ionization Chamber. An *ionization chamber* is a device that uses ionization of gases to measure radiation. They function similarly to Geiger counters in that radiation is measured by the amount of ionization in the gas. Ionization chambers are more accurate and their range is greater than Geiger counters. Ionization

SPECIFICATIONS FOR DOSIMETER #3700

Physical Characteristics

Detector Type	side window (30 mg/cm2) halogen quenched GM Tube (1-5114)
Detector Voltage Output	920 V
Readout Visual Audio	2.62 inch round meter Clicking type headset (optional)
Controls	4 position selector switch (off, × 1, × 10, × 100) Battery check switch
Power Batteries	2 - "D" cells (1.5 volt)
Connectors	phonojack for model 3020 headset
Mechanical Case Finish	Ruggedized aluminum body and cover Baked on enamel

Operating Characteristics

Radiation Detected	Beta, 0.8 MeV (53% efficient); Gamma and X-Ray—from 30 keV window open and 50 keV window closed
Accuracy	± 20%
Energy Dependence	± 30%, 80 keV to 1.25 MeV
Linearity	± 20%
Operating Ranges	0–0.5 mR/h; 0–5 mR/h; 0–50 mR/h 0–300 cpm; 0–3,000 cpm; 0–30,000 cpm
Exposure Rate Limit	1 R/h
Warm Up Time	None
Response Time	2 second time constant
Environmental Conditions Temperature Range Temperature Dependence Relative Humidity	14° to 122°F (-10°C to 50°C) ± 10% throughout operating range 20% to 95% (non-condensing)

Dimensions		
	Body	Probe (nickel plated)
Width	4.38″ (111 mm)	1.25″ (32 mm)
Length	.75″ (19 mm)	5.125″ (130 mm)
Height	4.25″ (108 mm)	.875″ (22 mm)
Net Weight	4 lbs. (1.8 Kg)	
Shipping Weight	6 lbs. (2.7 Kg)	

Accessories

Included Mounting	Carrying strap
Optional Check Source Audio	Model 3001, 5 μCi 137 Cs Model 3020 headset

Figure 5-5. Geiger counters show radiation levels through gauge readings or audible signals.

chambers are primarily used for continuous monitoring of radiation levels higher than a Geiger counter is capable of measuring.

Geiger counters are used in radiation fields up to 50 mrem/hour. For fields in the r/hour range, an ionization chamber is required. Ionization chambers are available for radiation fields up to 500 r/hour.

Scintillation Detector. A *scintillation detector* is a device that senses and measures radiation striking certain chemicals causing flashes of light (scintillations), which are converted to digital impulses. The digital impulses are counted and recorded by electronic equipment. Scintillation detectors are primarily used in laboratory settings for very accurate radiation readings. See Figure 5-6.

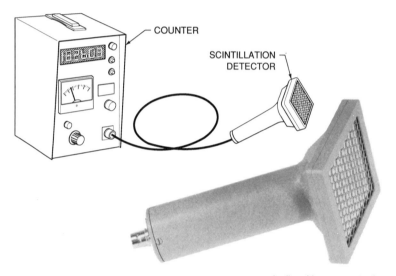

Ludlum Measurements, Inc.

Figure 5-6. Scintillation detectors record flashes of light (scintillations) caused by radiation.

Solid-state Radiation Detector. A *solid-state radiation detector* is a device that measures electrons released by radiation using semiconductors. The electrons released by radiation are collected and measured as a pulse or a continuous current of electricity. The use of semiconductors allows solid-state radiation detectors to be more compact and efficient than gas-filled radiation detectors. Solid-state detectors are used in radiation detection and monitoring equipment. The detector is located in a sensor that can be placed in a remote area. See Figure 5-7.

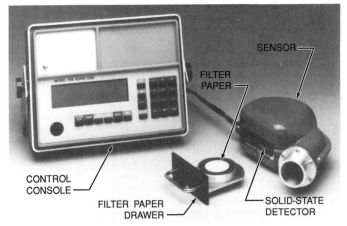

SENSOR

FILTER
PAPER

CONTROL
CONSOLE

FILTER PAPER
DRAWER

SOLID-STATE
DETECTOR

Victoreen, Inc.

Figure 5-7. Solid-state radiation detectors are smaller and more efficient than gas-filled radiation detectors.

Thermoluminescent Dosimeter. A *thermoluminescent dosimeter* (*TLD*) measures X rays and beta, gamma, and neutron radiation using lithium fluoride crystals. The crystals that are exposed to radiation give off light when heated. The amount of light emitted is measured to determine the amount of radiation exposure. The window in the TLD must face outward for proper dosage measurement. Filters on TLDs measure different types of radiation. TLD badges are worn between the neck and the waist. TLD rings are used to measure radiation dosages received by extremities (arms and fingers). See Figure 5-8.

TLD BADGE

TLD RING

Tech/Ops Landauer, Inc.

Figure 5-8. Thermoluminescent dosimeters measure whole-body doses of radiation.

TLDs are processed by heating and recording the measurement as a whole-body dose. Whole-body doses are listed on permanent records. The TLDs can be reused many times. Once heated during processing, however, the particular measurement is lost.

Direct-reading Pocket Dosimeter. A direct-reading pocket dosimeter is a gas-filled ion chamber that constantly monitors gamma radiation and X-ray levels. It is about the size of a fountain pen and can be read by the wearer while working. This allows the worker to monitor doses received on a continual basis. Readouts can be in roentgens or millirems. If the radiation exposure level approaches or exceeds maximum allowable limits, the worker can leave the area without overexposure. Direct-reading pocket dosimeters are commonly used in high radiation areas. Their range is less than TLDs and film badges. For this reason, TLDs and/or film badges are commonly used with them for proper protection. They are recharged by placing them into a dosimeter charger. See Figure 5-9. Readings are lost during recharging.

READING MEASUREMENT   RECHARGING

Dosimeter Corporation

Figure 5-9. Direct-reading pocket dosimeters are rechargeable portable radiation detectors that provide radiation monitoring on a continual basis.

Digital/Alarming Dosimeter. A digital/alarming dosimeter provides two ways of indicating radiation levels: digital readout or audible alarm. The digital/alarming dosimeter can be preset to detect a maximum radiation level. When a worker approaches or exceeds the maximum radiation level, the alarm will sound. In addition, it can also be used to identify areas with high radiation levels. The digital/alarming dosimeter is used for very high radiation levels or in emergency situations. See Figure 5-10.

Dosimeter Corporation

Figure 5-10. Digital/alarming dosimeters can be preset to indicate the presence of specific radiation levels.

Film Badge. A *film badge* contains film that is sensitive to radiation and is exposed in proportion to the amount of radiation received. After development, the darker the film badge is, the greater the dosage received. The film badge consists of a small piece of film that is wrapped to keep out light. It is mounted to a badge that can be clipped to clothing. The sensitivity of a film badge is similar to the sensitivity of a TLD. See Figure 5-11.

Film badges use filters similar to those in TLDs to measure different types of radiation. After development, film badges provide a permanent record of radiation dosages received. This allows nuclear power plant personnel to determine exactly how much radiation exposure occurred and to use the film badge to

Tennessee Valley Authority   *Tech/Ops Landauer, Inc.*

Figure 5-11. Personal film badges measure radiation levels for individual workers.

document the amount. Film badge analysis is used to determine permanent dose history.

Area Monitor. An *area monitor* is a device that monitors gamma radiation in specific locations. The area monitor is mounted on the wall in the area to be monitored. It is equipped with an audible alarm (bell) and a flashing red light that are activated when the gamma radiation level in the area exceeds a preset level. See Figure 5-12.

> **Warning:** Follow plant procedures when an area monitor is activated. Activation of this unit requires immediate evacuation of the area to a safer area of the plant (usually the control room).

Air Monitor. An *air monitor* continuously monitors air for airborne radioactivity at specific locations. Airborne radioactivity is caught on filter paper that moves past a radiation detector. The radiation detector then monitors the accumulation of radioactive material on the paper. The air monitor is also equipped with a flashing red light and an audible alarm that is activated when a preset level is reached. See Figure 5-13.

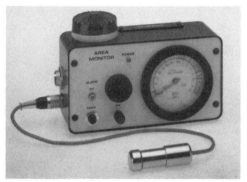

Dosimeter Corporation

Figure 5-12. Area monitors are stationary units that detect radiation in specific locations.

Eberline Instruments

Figure 5-13. Air monitors continuously monitor air for radiation levels.

Frisker. *Friskers* are radiation detection devices used for measuring radiation when leaving a controlled area. See Figure 5-14. Friskers can be used to check tools for radiation from a radiation

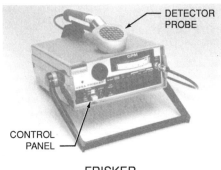

FRISKER

Victoreen, Inc.

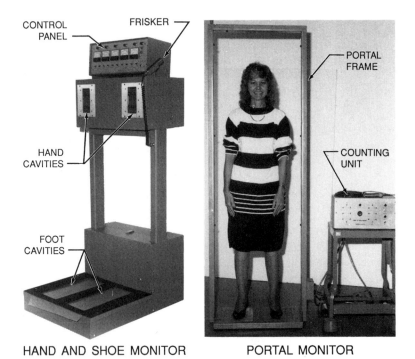

HAND AND SHOE MONITOR   PORTAL MONITOR

Ludlum Measurements, Inc.

Figure 5-14. Radiation levels of workers are measured using friskers, hand and shoe monitors, and portal monitors.

area. Friskers are also used to check body parts for areas of contamination. Alpha, beta, gamma, and X rays can be detected depending on the detector probe used. An audible alarm and scale indicate radiation levels.

Hand and Shoe Monitor. *Hand and shoe monitors* are radiation detection devices used for monitoring hands and shoes. Geiger-Muller (G-M) tubes are located at the foot cavities and the hand cavities of the monitor. Four G-M tubes are in each hand cavity. Two G-M tubes are in each foot cavity. Hand and shoe monitors can be designed to detect and measure alpha, beta, gamma, or X rays. An audible alarm and scale indicate radiation levels.

Portal Monitor. *Portal monitors* are radiation detection devices that surround the body of the worker. Eleven G-M detectors are located in the frame of a portal monitor. Seven detectors are located on the frame for monitoring the head and body. Four detectors are located in the portal monitor base for monitoring the feet. All workers leaving a controlled area are required to check through a portal monitor. If the alarm is triggered during checkout, the worker must report directly to health physics personnel for decontamination.

BIOLOGICAL EFFECTS

Biological effects from radiation can be somatic or genetic. *Somatic effects* cause direct damage to cell molecules in the body. *Genetic effects* are caused by radiation transferred from parent to offspring. Radiation exposure can be chronic or acute. *Chronic exposure* is small doses of radiation received over a long period of time. *Acute exposure* is a large dose of radiation received in a short time.

   In addition, biological effects from radiation can be prompt effects or delayed effects. *Prompt effects* of radiation appear shortly after exposure. *Delayed effects* of radiation may appear months or years after exposure.

Biological effects caused by radiation are determined by the part of the body exposed, the rate of exposure, the length of exposure, and the radiation type. Radiation can kill or damage cells. If enough cells die, the organ that the cells form will die. If crucial organs die, the organism will die.

Radiation exposure to the whole body exceeding 1000 rems over a brief period of minutes or hours is lethal. Lethal exposures occurred at the Hiroshima and Nagasaki bombings. A dose of 200 to 600 rems, delivered at one time to the whole body, will cause death in 10% to 80% of the cases within a one two-month period of time. A dose of 100,000 rems will cause death immediately in 100% of the cases. See Figure 5-15.

For low radiation doses it is cell damage, not cell death, that is harmful. A few dead cells can be replaced or repaired. However, cell damage can replicate itself and multiply. See Figure 5-16.

The type of cell damage resulting from radiation depends on the nature of the damaged cell. For example, if the damaged cell is bone or organ tissue, the damage is confined to that particular cell or organ (somatic effect). The most serious type of somatic effect of radiation is cancer. If cell damage occurs in reproductive cells, genetic damage through mutation can occur, transferring the radiation damage to offspring (genetic effect).

External exposure is biological effects caused by radiation exposure to external parts of the body. External exposure results in injury such as skin burns, lesions, loss of hair, and/or blood damage. Radiation sickness may also occur. *Radiation sickness* is any illness precipitated by exposure to radiation. The symptoms of radiation sickness include loss of appetite, nausea, hemorrhage, diarrhea, and decrease in red blood cell count.

Internal exposure is biological effects caused by the ingestion of radioactive materials. The radioactive material will migrate to body parts or organs depending on its chemical makeup. For example, uranium tends to collect in bone tissue and iodine collects in the thyroid gland.

BIOLOGICAL EFFECTS OF ACUTE RADIATION DOSES

Dose Rems	Symptoms/Effects	Latent Period	Characteristic Signs	Therapy	Convalescent Period	Incidence of Death	Time of Death	Cause of Death
0–25	no detectable effects (95–100%) minor blood changes (0–5%)	none 48 hours	none low white cell count	none none	none 1–2 weeks	0%		none
25–50	detectable changes in blood (95–100%)	24 hours	low white cell count low red cell count	blood transfusion	2–4 weeks	0%		none
50–100	detectable changes in blood (100%) nausea (0–5%)	12 hours 12 hours	low red cell count low white cell count vomiting	blood transfusion stomach relaxers	4–8 weeks 2–10 days	0%		none
100–200	nausea (5–50%) loss of body hair (0–10%)	3 hours 5–10 days	vomiting minor balding	stomach relaxers none	several weeks 6–12 months	0–10%	2–20 years	cancer
200–600	anemia nausea loss of body hair (10–50%) heart palpitations	1–12 days 2 hours 2–5 days 1 hour–12 months	leukopenia purpura hemorrhage severe vomiting balding rapid pulse	blood transfusion antibiotics stomach relaxers reassurance	1–12 months	10–80%	1–2 months	hemorrhage infection heart failure cancer/leukemia
600–1000	severe anemia nausea-diarrhea severe loss of body hair heart palpitations/heart attack	1 hour 1 hour 2–3 days 1 hour	leukopenia purpura hemorrhage intestinal discharge severe balding rapid pulse heart stoppage	blood transfusion blood coagulants bone marrow transplants intravenous feeding adrenalin	extremely long	80–100%	2 days–2 months	infection hemorrhage heart failure kidney failure respiratory failure cancer/leukemia
1000–100,000	pernicious anemia nervous system failure nausea-diarrhea heart, respiratory, excretory failure	1 hour 30 minutes 30 minutes 1 hour–30 minutes	leukopenia purpura hemorrhage convulsions tremor-lethargy	blood transfusion bone marrow transplants sedatives	extremely long	99–100%	1 hour–2 weeks	infection, hemorrhage, heart failure, kidney failure respiratory failure coma brain failure
100,000	complete malfunction	none	cell wall rupture	none	none	100%	instantly	molecular death

Figure 5-15. The biological effects of acute radiation doses range from no effect to immediate death.

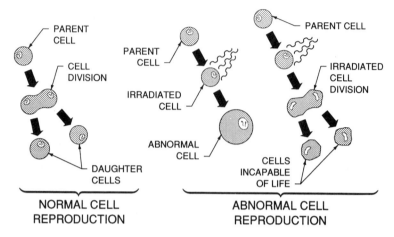

Figure 5-16. Irradiated cells can multiply, creating cells incapable of life.

Lethal Dose

Lethal dose (*LD*) is the term used to describe the deadly effects of radiation exposure. An abbreviation is used to describe the dose equivalent to kill a certain percentage of people exposed. For example, LD50 is the dose equivalent to kill 50% of the people exposed. This lethal dose is in the range of 300 to 500 rem.

While the effects of radiation exposure can cause cancer or genetic mutation, the relationship between the dose and the incidence of cancer and genetic mutation is unclear. This is especially true in doses below 1 rem and the incidence of resulting cancers or genetic mutations. For both types of damage, the latent period is long. A *latent period* is the time between radiation exposure and its effect. The latent period of cancer can be 25 years or more. One or more generations may pass before genetic damage appears. In both cases, other possible causes such as other chemical carcinogens can make it difficult or impossible to trace the origin of the cancer. Statistics compiled regarding the nuclear power industry indicate fewer health risks and a good safety record when compared to other occupations and industries.

NUCLEAR ACCIDENTS

The nuclear industry has had an excellent safety record. However, three major accidents have occurred in the past three decades. The first nuclear industry accident occurred on October 10, 1957, at the Windscale nuclear weapons plant located on the east coast of England. Windscale produced plutonium for nuclear weapons. The reactor used graphite as a moderator. A routine procedure of heating the graphite was performed to reduce the normal swelling of graphite bombarded by neutrons in a nuclear reactor. In this instance, the graphite required heating a second time because the first heating did not produce the desired results. During the second heating, the graphite temperatures continued to rise abnormally. The graphite ignited, and efforts to put out the fire were unsuccessful. Finally, after using large amounts of water, the fire was struck. In the process of putting out the fire, iodine 131 (^{131}I), cesium 137 (^{137}Cs), and other radioactive isotopes were released into the atmosphere. The extent of health hazards to persons and livestock near the plant has not been fully determined.

The second major nuclear accident occurred on March 28, 1979, at the Three Mile Island nuclear power plant near Harrisburg, Pennsylvania. The accident, which occurred during normal operation, was the result of a series of mechanical failures and operator errors. The problem began when cooling water supplied to the reactor core was cut off. As pressure in the reactor core increased, a safety valve opened and the reactor shut down automatically. In the process of trying to determine what had occurred, plant operators thought the safety valve had closed. However, the safety valve remained open, flooding the core. Responding to preventing the flow of water from the safety valve from flooding the reactor core, plant operators manually shut off the emergency core cooling system. Two hours passed before this mistake was discovered. By the time it was discovered, the reactor core began to boil. A steam bubble formed, which exposed part of the reactor core. Radioactive material from the reactor core escaped and entered the containment building. Some of this radioactive material mixed with coolant water through the open safety valve into an auxiliary building housing water storage tanks. A small amount of radioactive material was released through a charcoal filter in a stack into the atmosphere.

A comprehensive investigation of the Three Mile Island accident concluded that many factors contributed to the accident. These include faulty instrumentation and equipment, and the inability of plant operators to quickly assess the situation and read complex display panels in the control room. This led to modifications in plant operator training, plant equipment and procedures, and control room design.

The third major accident occurred on April 26, 1986, at a nuclear power plant facility consisting of four reactors at Chernobyl near Kiev in the Soviet Union. Like Windscale, the reactors used graphite as a moderator and water as a coolant. The design of these plants, identified as RBMK-1000, is significantly different from plant designs used in the United States. These design differences contributed to the severity of the accident. For example, all plants in the United States must have containment to serve as a barrier in the event of a problem with the reactor. In addition, plants in the United States will automatically reduce power level in the reactor in the event of a loss of cooling water. The opposite occurs in an RBMK-1000 reactor. If cooling water turns to steam, the power level in the reactor increases.

During a scheduled shutdown, technicians at the Chernobyl facility began a test that required the reactor to operate at a low power level. RBMK-1000 reactors are difficult to control at low power levels. In addition, to obtain this low power level, several safety systems designed to shut down the reactor in the event of an accident were deactivated. As the test was conducted, the reactor had power level fluctuations and became hotter. Efforts to control the power level increase were unsuccessful. Steam pressure rose to the point where there was an explosion caused by a sudden release of steam. This blew the top off the reactor building. A second explosion occurred from the ignition of hydrogen from fuel rods. This caused graphite in the reactor core to catch on fire. The fire burned for several days, during which time large amounts of radioactive isotopes were released into the atmosphere. Large amounts of sand, lead, clay, dolomite, and boron were used to extinguish the fire. Over 100,000 people in areas surrounding the plant were evacuated. The radiation exposure and burns resulted in 32 deaths and hundreds of people injured and hospitalized. The reactor core at reactor four has been sealed in concrete to prevent any additional radiation leaks. The Chernobyl accident has resulted in new regulations in the Soviet Union for plant operation and training procedures, notification requirements of nuclear accidents, and assistance provided by other nations.

PUBLIC CONCERNS

The perception of the public toward the safety of nuclear power has a great deal to do with the future of nuclear power. Federal, state, and local regulations ensure the safety in the operation of nuclear power plants. In addition, the safety record of nuclear power plants in the United States is exemplary. The Three Mile Island accident caused concern regarding the future of nuclear power. In this most serious accident in the United States nuclear industry, it was determined by a 1985 Pennsylvania Department of Health study that there has been no significant increase in cancer rates in areas that were downwind of the plant.

Nuclear power plants will continue to be an integral part of the energy required by the United States economy in the future. See Figure 5-17. The increasing role of nuclear power plants is illustrated by the following statistics. Electricity produced using nuclear energy has
- doubled in capacity since 1979,
- surpassed oil as a source of electricity in 1980,
- surpassed natural gas as a source of electricity in 1983, and
- surpassed hydroelectric as a source of electricity in 1984.

In addition to these statistics, concerns have been raised regarding the effects of use of fossil fuels on the environment, dependence on oil supplied from foreign countries, and meeting the increasing demand for electricity. Effects of fossil fuels on the environment include the problems of acid rain and the greenhouse effect. The effects of dependence on oil supplied from foreign countries was at its peak during the 1973-74 Arab oil embargo. Although interruption to oil supplied from foreign countries has not occurred recently, changes in global political structures and growing United States trade deficits have raised new concerns. The demand for electricity has increased from 27% of all energy use to 37% since 1973. This demand has resulted in brownouts in the northeast and mid-Atlantic states during peak usage in 1988 and 1989.

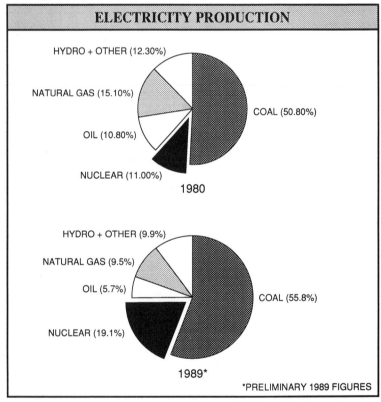

U.S. Council for Energy Awareness

Figure 5-17. The use of nuclear power to produce electricity is increasing.

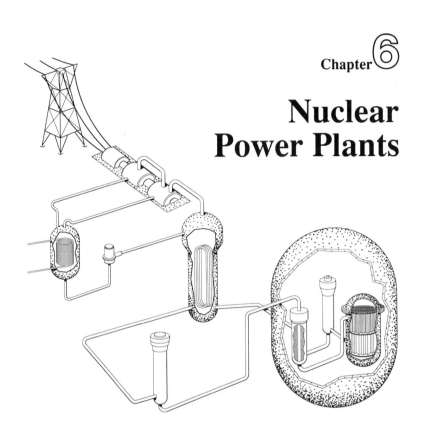

Chapter 6

Nuclear Power Plants

The major difference between fossil fuel power plants and nuclear power plants is the fuel used to generate heat. Typical nuclear power plant facilities consist of a reactor building, turbine generator building, radwaste building, control room, and cooling water provisions. Each nuclear reactor is housed in a separate reactor building.

Light water reactors (LWRs) are the most commonly used reactor type in the United States. Water is used as a moderator and a medium to transfer heat from the reactor core in LWRs. High temperature gas-cooled reactors (HTGRs) use helium, Canadian deuterium-uranium (CANDU) reactors use deuterium oxide (heavy water), and liquid metal fast breeder reactors (LMFBRs) use liquid sodium.

REACTORS

Nuclear power plants throughout the world use different reactor types. Reactor types include light water reactors (LWRs), high temperature gas-cooled reactors (HTGRs), Canadian deuterium-uranium (CANDU) reactors, and liquid metal fast breeder reactors (LMFBRs). The most commonly used reactor type in nuclear power plants is the LWR. The role of reactors in nuclear power plants is the same regardless of the reactor type. The primary difference between reactor types is how heat is generated.

Light Water Reactors

Light water reactors (LWRs) use water as a moderator and a medium to transfer heat from the reactor core. The two types of light water reactors are boiling water reactors (BWRs) and pressurized water reactors (PWRs). In BWRs, water flows through coils located around the reactor core. The heat turns the water into steam. The steam is piped directly to the steam turbines.

In PWRs, water passing around the reactor core is pressurized to approximately 1000 psi. Pressure increases the temperature at which the water will turn to steam. The heated water is piped to the steam generator. Water in a second loop is piped to the steam generator where it is turned to steam. The steam is then piped to the steam turbines. See Figure 6-1.

High Temperature Gas-cooled Reactors

High temperature gas-cooled reactors (HTGRs) are less commonly used than light water reactors. In an HTGR, helium is used for cooling and for transferring heat from the reactor core to the steam generator. Steam is produced from a secondary loop passing through the steam generator. The steam is then piped to the steam turbines. See Figure 6-2.

Fuel used in an HTGR is a mixture of uranium 235 (^{235}U) and thorium 232 (^{232}Th) carbide. The only HTGR plant built in the United States was at Fort St. Vrain, Colorado. This plant,

BWR PLANT

Tennessee Valley Authority

PWR PLANT

U.S. Council for Energy Awareness

Figure 6-1. The two types of light water reactors (LWRs) are boiling water reactors (BWRs) and pressurized water reactors (PWRs).

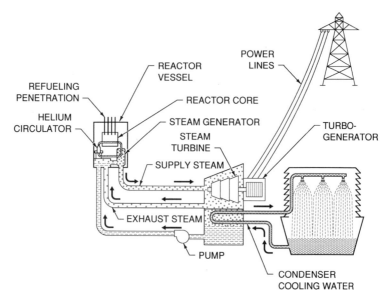

Public Service Company of Colorado, Fort St. Vrain

Figure 6-2. No high temperature gas-cooled reactor (HTGR) plants are currently operating in the United States.

which was announced as the nation's first HTGR demonstration plant in 1965, generated its first electricity in 1976. It had experienced shutdowns due to moisture in the reactor core and is no longer in operation. Several HTGR plants are currently operating in the United Kingdom.

Canadian Deuterium-Uranium Reactors

Canadian deuterium-uranium (CANDU) reactors use natural uranium fuel that has not been enriched above the natural content of 0.7% ^{235}U, and 99.3% ^{238}U. Normally, this fuel will not sustain the fission process required for a chain reaction. However, a mixture of deuterium oxide (heavy water), when used as a moderator and a coolant, will sustain a chain reaction using less expensive natural uranium. The heavy water under pressure transfers heat to the steam generator. Water in a second loop is heated and converted into steam for use in the steam turbines. The high cost of heavy water used in CANDU reactors is offset by the savings gained by using natural uranium. See Figure 6-3.

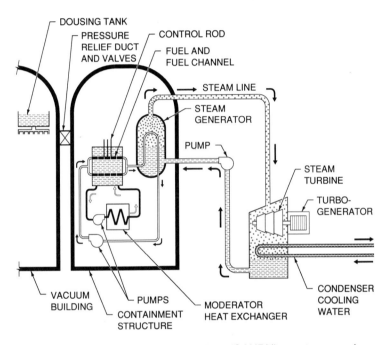

Figure 6-3. Canadian deuterium-uranium (CANDU) reactors use heavy water and less expensive natural uranium to sustain the fission process.

Liquid Metal Fast Breeder Reactors

Liquid metal fast breeder reactors (LMFBRs) produce more fuel than they consume. Uranium 235 is surrounded by uranium 238 (^{238}U). During the nuclear reaction, the ^{238}U is bombarded with fast-moving neutrons. Fast-moving neutrons are usually slowed using a moderator. When bombarded with the fast-moving neutrons, ^{238}U is converted to plutonium 239 (^{239}Pu). In optimum conditions, for every four atoms of fuel consumed, five atoms of fuel are produced. Liquid sodium is used to transfer heat from the reactor core to a second loop. The second loop also contains liquid sodium and transfers heat to the steam generator. A third loop directs water to the steam generator and steam to the steam turbine. Plutonium 239 that is produced in a breeder reactor can be reprocessed into additional fuel.

Funding for the Clinch River Breeder Reactor was discontinued by the U.S. Congress in 1983 and the project was not completed. There are currently no commercial breeder reactors operating in the United States. However, France, Japan, the Federal Republic of Germany (West Germany), and the Soviet Union have LMFBRs. See Figure 6-4. Research in LMFBR technology still is being conducted by the Department of Energy.

U.S. Council for Energy Awareness

Figure 6-4. Liquid metal fast breeder reactors (LMFBRs) produce ^{239}Pu, which can be reprocessed into additional fuel.

NUCLEAR POWER PLANT FACILITIES

Nuclear power plant facilities consist of components that are similar to those required in fossil fuel power plants. The primary difference is the fuel used to generate heat. In fossil fuel power plants, heat from the combustion of natural gas, coal, or other fossil fuels is used to produce heat required to generate steam. In nuclear power plants, heat from nuclear fission is used to produce heat required to generate steam.

Specialized facilities and equipment are required for control and safety in nuclear power plants. The physical location and configuration of these facilities vary from plant to plant. Nuclear power plants commonly require 200 to 300 acres of land and are usually located near water sources such as a lake or river. Water is used for cooling in the nuclear reaction and steam generation processes.

A typical nuclear power plant consists of primary facilities including a reactor building, turbine generator building, radwaste building, control room, and cooling water provisions. See Figure 6-5. Other support facilities such as pumphouses and service buildings and transformer yard are located at nuclear power plant facilities.

Reactor Building

Each nuclear reactor is housed in a separate reactor building. The reactor building houses the containment structure, which serves as shielding to the reactor. The reactor building also protects the containment structure from the weather. The containment structure has an outer shell of approximately 3′ of reinforced concrete. A ¾″ steel liner is used in the inside of the containment structure. The containment structure is designed to protect radiation from leaking out and prevent damage caused from severe weather or military assault. See Figure 6-6.

Turbine Generator Building

The turbine generator building houses the steam turbines and generators used to generate electricity. Because steam turbines are directly linked to generators in power plants, they are sometimes referred to as turbo-generators. In addition, steam turbines are often referred to as prime movers. *Prime movers* are mechanical devices used to drive generators. To generate more electrical energy requires more mechanical energy from the prime

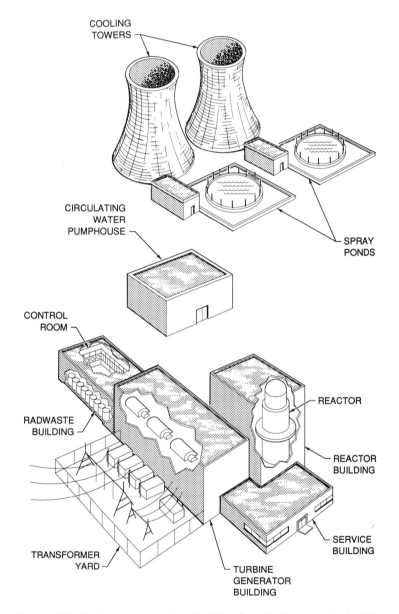

Figure 6-5. Nuclear power plant facilities include the reactor building, turbine generator building, radwaste building, control room, cooling water provisions, and other buildings as required.

Figure 6-6. Each nuclear reactor is housed in a separate reactor building, which includes the containment structure shielding the reactor.

mover. Steam turbines in the turbine generator building convert heat energy in the steam into kinetic energy to drive the generators. Steam turbines in power plants use hydraulic governors for accurate speed control. Fluctuation in the speed of the steam turbine would affect the cycles being generated. The hydraulic governors also prevent a steam turbine from exceeding safe operating speed if the load suddenly changes.

Steam produced is piped to the steam turbine. Steam turbines used in power plants contain many sets of wheels that can be divided into stages. In the steam turbine, steam at a high velocity is directed through wheels and stationary blades of each stage. See Figure 6-7. The force of the steam causes the wheels, which are mounted on a common axle, to rotate rapidly. As the steam passes from one stage to the next stage, the volume of the steam is increased. Each stage of the steam turbine is larger than the previous stage to maximize the efficiency of the expanding steam.

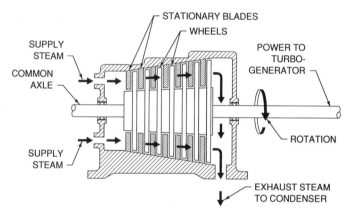

Figure 6-7. The volume of steam in steam turbines is increased by a series of wheels and stationary blades that cause the common axle, connected to the turbo-generator, to rotate rapidly.

After passing through the steam turbine, the steam is referred to as exhaust steam. The exhaust steam is directed to the condenser. Steam passes around tubes containing cooling water in the condenser. The steam condenses to water (condensate) as it is cooled. The condensate is then returned for reuse in the system. Steam and cooling water do not mix in the condenser. Condensers are classified as single-pass or multipass depending on how many passes the cooling water makes before leaving the condenser.

The steam turbine is directly connected to the generator. The two types of generators are alternating current (AC) and direct current (DC) generators. Virtually all generators at power plants are AC generators. Alternating current generated at a typical power plant has 18,000 to 22,000 V. This voltage can be increased or decreased as necessary using transformers. The voltage is increased to as much as 765,000 V for sending electricity long distances over power lines.

Radwaste Building

The radwaste building houses the equipment required for processing radioactive waste produced in the nuclear power plant. Tanks are used for collection of radioactive waste before processing. Equipment is used for containers and loading and shipping radioactive waste material.

Control Room

The control room can be housed in a separate building or within another building at the nuclear power plant. The control room contains instrumentation used to monitor and control the operation of the nuclear power plant. This includes main and auxiliary control panels and emergency switchgear. Emergency switchgear is used to provide on-site emergency power in the event of an off-site power failure. See Figure 6-8.

Cooling Water Provisions

Cooling water provisions are required for fossil fuel and nuclear power plants. Consequently, most power plants are located near lakes, rivers, or other large bodies of water. Cooling water used in power plants cools and condenses steam after use in the steam turbine. The temperature of the cooling water is raised during this process. Cooling towers are used to lower the temperature of the cooling water used. After the cooling water is cooled to the proper temperature, it is returned to its original source (for example, a lake or river). Condensate in the nuclear power plant

Figure 6-8. The control room contains instrumentation to monitor nuclear power plant operations.

does not come in contact with the nuclear reactor or other radioactive equipment.

An adequate supply of cooling water is required for condensing steam. Up to 50 pounds of water may be required to condense 1 pound of steam. This amount varies with ambient air temperature and cooling water temperature. The cooling water system may use either a cooling pond or spray pond or a cooling tower. The cooling pond is simpler, requires less initial cost, but is not as effective as a cooling tower. Both the cooling pond and cooling tower rely on the evaporation of a portion of the water to achieve the necessary cooling effect. From 5% to 10% of the water is lost through evaporation in the cooling process. This loss of water must be recovered to maintain the proper water level in the system.

Cooling ponds must be large enough to cool water to the temperature desired by the surface contact of the water to the air. This may require a pond too large in area to be practical. Spray ponds are cooling ponds in which water is sprayed. Spray ponds increase the surface area of water in contact with the air by breaking up the water into a fine spray. A spray pond can be smaller than a cooling pond with similar cooling efficiency.

Cooling towers provide the most efficient method of cooling water. The water is discharged to the top of the tower and broken into fine droplets that fall into a tank. Water releases heat to the air by evaporation, convection, and radiation. Evaporation absorbs 75% to 85% of the heat. The heat that must be extracted in the cooling tower is equal to the latent heat absorbed by the cooling water while condensing the steam in the condenser. In addition, the cooling tower must evaporate a quantity of water equaling approximately 80% of the weight of the steam passing through the turbo-generator. The water being cooled may be lowered in temperature by 30°F to 40°F. See Figure 6-9.

The flow rate of water used for cooling is measured in gallons per minute (gpm). The quantity of water required for condensing steam in steam turbines varies depending on the amount of steam to be condensed, temperature of the steam at the condenser, temperature of the cooling water available, and type of condenser used.

Zurn Construction, Inc./Balcke Dürr, AG

Figure 6-9. Cooling towers are the most efficient method of cooling water.

POWER DISTRIBUTION

Power distribution is required to transmit electrical power from the nuclear power plant to points of use for industrial, commercial, and residential applications. Power distribution systems are designed for efficiency and reliability. This requires careful planning and coordination to prevent a failure of the system. Failure of the system can result in blackouts. A *blackout* is the loss of electric power caused by an interruption of service from the supplier. Interruptions in service are commonly caused by the weather or accidents. Blackouts can cause serious health, safety, and economic repercussions.

Compensation for increased power demands such as those that may be caused by extraordinary weather conditions (either extreme cold or extreme heat) is also required to prevent brownouts. A *brownout* is the reduction of voltage from the supplier as a measure to prevent a blackout. Brownouts can cause damage or reduce the efficiency of electrical equipment.

Nuclear power plants generate electricity using steam turbines to drive alternating current generators. Alternating current is required for transmitting electricity over great distances. Alternating current can be rectified later to direct current as required. Transformers are used to increase or decrease voltage as required.

Electricity generated at the nuclear power plant is routed to step-up transformers. A *step-up transformer* "steps up" (increases) voltage for long-distance transmission of electricity over power (transmission) lines. Power lines and towers are used to transmit high-voltage electricity rather than underground cables because of cost. Increasing the voltage allows the use of smaller gauge wires to transmit electricity through power lines. For example, a generated voltage of 18,000 V may be stepped up to 200,000 V by a step-up transformer. The 200,000 V can then be carried over smaller gauge power lines. Increasing the voltage decreases the amount of current required to transmit electricity and increases transmission efficiency. The higher the transmission voltage is, the lower the transmission current will be.

High-voltage electricity from the power plant is carried over towers and power lines to transmission substations and large industrial plants before transmission to distribution substations where it is routed to industrial, commercial, and residential users. See Figure 6-10.

Substations provide switching for various distribution needs, increase the voltage of electricity to be transmitted farther, and reduce voltage for industrial, commercial, and residential use. Substations may be housed in buildings or located in open areas. Fencing is placed around the substation. Generally, substations are located for maximum efficiency and close proximity to the point of use.

Transmission Substations

High-voltage electricity is transmitted from the power plant through power lines to transmission substations. *Transmission substations* are electrical facilities that provide switching for various needs and decrease the high voltage of electricity from power plants for lower voltage power lines to distribution substations and heavy industry. For example, the voltage may be reduced from the 200,000 V required for long-distance transmission from the power plant to 22,000 V for shorter distance transmission to distribution substations or to heavy industry. Power line voltage of 22,000 V is used to supply power to large industrial plants.

Distribution Substations

Distribution substations are electrical facilities that provide switching for various needs and decrease the voltage of electricity from transmission substations to lower voltage electricity, which is distributed to step-down transformers of industrial, commercial, and residential users. *Step-down transformers* "step down" or reduce voltage for distribution to consumers of electricty. Distribution power lines transmit electricity from distribution substations to the transformers that make the final voltage step-down for end use.

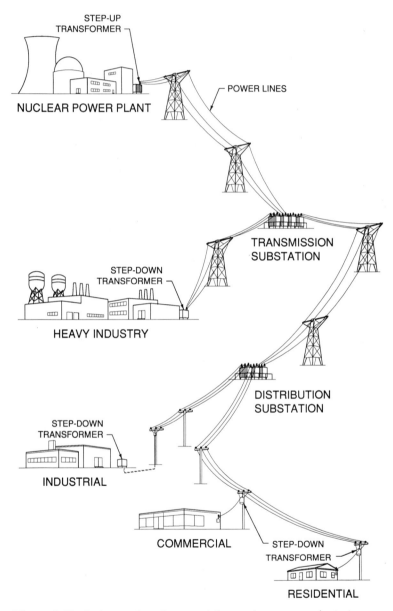

Figure 6-10. A step-up transformer at the nuclear power plant steps up voltage for long distance transmission of electricity. Step-down transformers step down voltage for end users.

Transformers

The three major types of transformers are overhead transformers, pad-mounted transformers, and underground vault transformers. *Overhead transformers* are mounted on the pole and provide overhead electrical service to buildings. *Pad-mounted transformers* are placed on a concrete pad and provide lateral (underground) electrical service to buildings. Pad-mounted transformers are enclosed by a weatherproof, tamperproof enclosure to prevent exposure to live parts. *Underground vault transformers* are in vaults beneath grade level and provide electrical service to buildings where all lines are underground. Access to underground vault transformers is provided by grating installed at street level.

Overhead, pad-mounted, and underground vault transformers reduce distribution line voltage from the distribution substation to utilization voltage as required for specific building uses. Distribution line voltage is usually 4000 V to 13,800 V.

Utilization Voltages

Electricity of various voltages is transmitted from transmission substations to heavy industry and to distribution substations. Distribution substations transmit electricity to industrial, commercial, and residential users. Common utilization voltages are

- Industrial   14,400/24,900 V, 3φ
                       12,470/21,600 V, 3φ
                       7200/12,470 V, 3φ
                       2400/4160 V, 3φ
                       2400 V, 3φ

- Commercial   277/480 V, 3φ
                       120/208 V, 3φ
                       120/240 V, 3φ
                       120/240 V, 1φ

- Residential   120/240 V, 1φ

	RADIATION PROTECTION REQUIREMENTS
COVERAGE	___ CONTINUOUS ___ PERIODIC ✔ INITIAL SURV. ONLY ___ SEE SPECIAL INSTR. ___ SURVEY FREQUENCY ___
MONITORING	✔ TLD ✔ 0-500 mR DOSI *OR* ✔ HIGH-RANGE DOSI ___ SEE SPECIAL INSTR.
BODY	___ LAB COAT ✔ 1 PAIR COVERALLS ___ 2 PAIR COVERALLS ___ WATERPROOF ___ SEE SPECIAL INSTR.
HEAD	✔ HOOD ___ WATERPROOF COVER ___ EYE SHIELD ___ CAP ___ SEE SPECIAL INSTR.
HAND	✔ COTTON LINERS ___ COTTON WORK GLOVES ✔ RUBBER GLOVES _1_ PR. ___ SEE SPECIAL INSTR.
FEET	✔ RUBBER OVERSHOES ___ RUBBER BOOTS ✔ PLASTIC BOOTIES ___ SEE SPECIAL INSTR.
RESPIRATORY	___ FULL FACE NP ___ SUPPLIED AIR ___ SCBA ___ AIR HOOD ___ SEE SPECIAL INSTR.

Chapter 7

Nuclear Power Plant Procedures

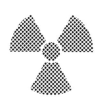

SLOT NO.	AUTH. EXP.			RADIATION EXPOSURE CARD						AVAILABLE DOSE FROM PREVIOUS CARD ___	
	1000		WP-1094 R4 (3-85)								
			DATE	RWP NO.	TASK (MWR NO.) (R-S-P-E)	TIME		DOSIMETER READING		NET DOSE (MR)	AVAILABLE REMAINING DOSE	REZERO
						IN	OUT	IN	OUT			
yes	00 7.5.11		1-8-90	2-90-1	AX 1184	0910	1430	5	35	30	970	
			1-9-90	2-90-4	AY 1201	0803	1450	35	0			

Security in nuclear power plants is required to protect workers and ensure efficient plant operation. Radiation exposure limits are constantly monitored and documented to ensure the safety of workers. Radiation work permit (RWP) requirements must be followed to minimize exposure.

In the event of an emergency in a nuclear power plant, emergency action levels are declared that require specific responses. The four emergency action levels are notification, alert, site emergency, and general emergency. A plant procedures manual specifies and standardizes required procedures to follow in nuclear power plants.

SECURITY

Security is required to ensure the safe and efficient operation of nuclear power plants. Specific security procedures vary slightly from plant to plant. It is the responsibility of the worker to follow specific plant security policies and regulations. As in other industrial plants, alcoholic beverages, nonprescription drugs, explosives, and firearms are not permitted in nuclear power plants. Cameras are allowed for specific use only with proper authorization.

Personnel authorized in the nuclear power plant are identified with photo identification badges. Plant security personnel issue photo identification badges as required. Each badge has a number and an access color, number, and/or letter code that authorizes access to certain areas of the nuclear power plant. This readily identifies the worker and function of the worker inside the plant. Visitors are generally required to have an authorized escort present at all times. Keycards with an electronic code allow access to authorized areas programmed into the security system.

Access to specific areas of the plant change according to the radiation level and function of the area. For example, a keycard reader could be programmed to allow an electrician access to an area in the radwaste building for a day. The keycard reader is then reprogrammed to deny access to the same area when the job is completed.

A TLD and film badge are issued as required. Some nuclear power plants have film badges or TLDs built into the identification badge. Personnel must proceed through an explosive detector. The explosive detector will activate if explosives are present. The explosive detector will also activate if there is excessive body movement or strong odors from paint, smoke, after-shave, cologne, perfume, etc. A search is required whenever the explosive detector is activated. Hand-held items must be sent through the X-ray machine. TLDs must not be sent through the X-ray machine.

Personnel then proceed through the metal detector. Metal detectors, like those used at airports, detect the presence of metal.

This prevents the possibility of a person possessing a concealed weapon from entering the nuclear power plant. Metal detectors use magnetic fields to sense metal objects. As metal passes through the magnetic field, a different magnetic field is created. This identifies a metal object. The sensitivity of a metal detector can be set to allow coins, belt buckles, and other small masses of metal to pass.

Keycards provide employee access to designated plant areas and are required for passage through area entries. Only one person may enter an area at a time. The keycard is read by the keycard reader and will unlock the door if access is permitted. The door is programmed to remain open long enough for one person to pass through the entry. Exit from an area requiring a keycard follows a similar procedure as entry to an area.

All security problems and violations must be reported to plant security or the supervisor in charge of the area. Security rules and regulations are for the safety and protection of all workers in the plant. Violation of any security rules or regulations may result in disciplinary action.

RADIATION EXPOSURE LIMIT

The federal radiation exposure limit, as determined by the DOE and enforced by the NRC for persons working in nuclear power plants, is a maximum of 5000 mrem per year with no more than 3000 mrem during any calendar quarter. For example, if a worker in a nuclear power plant received 1400 mrem during the first quarter of a calendar year, a total of 3600 mrem could not be exceeded during the remaining three quarters. No more than 3000 mrem could be exceeded in any one quarter. Individual nuclear power plants establish their own limits, which are usually lower than federal limits. See Figure 7-1.

Radiation doses received by workers are documented on a radiation exposure card. A typical radiation exposure card contains spaces for the employee's name, slot number in the card rack, identifying number (Social Security number, employee number, etc.), authorized exposure TLD number, and dosimeter number.

TYPE OF EXPOSURE	GUIDE VALUE
whole body; head and trunk; blood forming organs; lens of eye; or gonads	1250 millirems per quarter; 5000 millirems per year; up to 3000 mrems permitted in a quarter provided accumulated occupational dose to whole body does not exceed 5000 mrems × (age-18)
skin of whole body	7500 millirems per quarter
hands, forearms, feet and ankles	18,750 millirems per quarter

Figure 7-1. Code of Federal Regulations, Title 10, Part 20 details guidelines for radiation exposure.

Information is recorded on the radiation exposure card each time a worker leaves a controlled area. For example, on 1-11-90 while working under radiation work permit (RWP) number 2-90-1, J. Jones received 65 mrem of radiation from 7:30 to 11:10 AM while performing task AX 1195. The 65 mrem received reduced the available remaining dose to 760 mrem. See Figure 7-2.

Radiation exposure cards are maintained for varying amounts of time based on plant standards and requirements. Health physics is responsible for maintaining records related to radiation exposure.

SLOT NO.	AUTH. EXP.		RADIATION EXPOSURE CARD WP-1094 R4 (3-85)								AVAILABLE DOSE FROM PREVIOUS CARD	
			DATE	RWP NO.	TASK (MWR NO.) (R-S-P-E)	TIME		DOSIMETER READING		NET DOSE (MR)	AVAILABLE REMAINING DOSE	REZERO
						IN	OUT	IN	OUT			
	1000	007151!	1-8-90	2-90-1	AX 1184	0910	1430	5	35	30	970	
			1-9-90	2-90-4	AY 1201	0803	1450	35	95	60	910	
	2000100		1-10-90	2-90-1	AX 1192	0835	1115	95	180	85	825	RW
			1-11-90	2-90-1	AX 1195	0730	1110	0	65	65	760	
			1-11-90	2-90-4	AY 1210	1215	1530	65	65	0	760	

NAME (PRINT) J. Jones SOCIAL SECURITY NO. T.L.D. NO. DOSIMETER NO.

Washington Public Power Supply System

Figure 7-2. Radiation exposure cards are used to document radiation exposure received to prevent exceeding maximum dosages established by the nuclear power plant.

In addition to providing upper limits on radiation exposure, the NRC also requires licensees to minimize exposure to radiation and radioactive materials as low as reasonably achievable (ALARA). The standard of ALARA dictates that every activity at a nuclear power plant should be planned so as to minimize unnecessary exposure to the worker and to the worker population as a whole. To accomplish ALARA, nuclear power plants must maintain a staff of health physics personnel to monitor each task occurring in a radiation area. Additionally, a safety committee must review and approve plant procedures and practices.

While federal and local authorities set and enforce standards, minimizing radiation exposure is the responsibility of the individual worker. Workers can minimize their exposure by practicing the following general rules:

- Wear protective equipment as required.
- Follow specified emergency procedures.
- Monitor personal radiation dose status to prevent exceeding dose limits.
- Do not enter posted areas unless authorized.
- Adhere to all radiation work permit (RWP) requirements. Perform appropriate personal survey as required when exiting a controlled area.
- Minimize time spent in radiological areas. Perform all tasks in as low a radiation area as possible.
- Follow all plant rules, regulations, and procedures.

Health Physics

Health physics personnel are responsible for providing radiologically safe working conditions in nuclear power plants according to state and federal regulations. The role of health physics in nuclear power plants originated in the Atomic Energy Act of 1954. The NRC, through its Operational Inspection Program, performs routine appraisals of health physics procedures in nuclear power plants. Failure to comply with NRC requirements can result in suspension of a nuclear power plant license.

Health physics personnel are present to provide maximum safety for all workers in a nuclear power plant. They are readily accessible to all workers. Questions regarding radioactivity, dosimetry, contamination, noncompliance, and injuries should be directed to health physics personnel.

Controlled Areas

Controlled areas are nuclear power plant areas in which radioactive materials or radiation may exist. According to the Department of Energy (DOE) order DOE 5480.11,

> Controlled Area. The access to any controlled area where radioactive materials or elevated radiation fields may be present shall be clearly and conspicuously posted as a controlled area. The type of sign used may be selected by the contractor with the approval of the field organization to avoid conflict with local security requirements.

Radiological Areas

Nuclear power plants can have three different types of radiological areas: radiation area, high radiation area, and very high radiation area. Each of these areas is posted with the distinctive yellow and magenta radiation warning sign. See Figure 7-3. According to the Department of Energy (DOE) order DOE 5480.11,

> Posting for External Radiation. The access to any area where an individual can at anytime during normal operations receive a dose equivalent greater than 5 mrem in 1 hour at 30 centimeters from the radiation source or any surface through which radiation penetrates shall be posted as below. In addition, the anticipated dose rate or range of dose rates shall be included on or in conjunction with each of the signs, as appropriate.
> "Radiation Area" for any area within a controlled area where an individual can receive a dose equivalent greater than 5 mrem but less than 100 mrem in 1 hr at 30 cm from the radiation source or from any surface through which the radiation penetrates,
> "High Radiation Area" for any area within a controlled area where an individual can receive a dose equivalent of 100 mrem or greater but less than 5 rem in 1 hr at

30 cm from the radiation source or from any surface through which the radiation penetrates, and

"Very High Radiation Area" for any area within a controlled area where an individual can receive a dose of 5 rem or greater in 1 hour at 30 cm from the radiation source or from any surface through which the radiation penetrates.

Posting for Airborne Radioactive Material. The access to any area where airborne radioactive material concentrations greater than 1/10 of the derived air concentrations

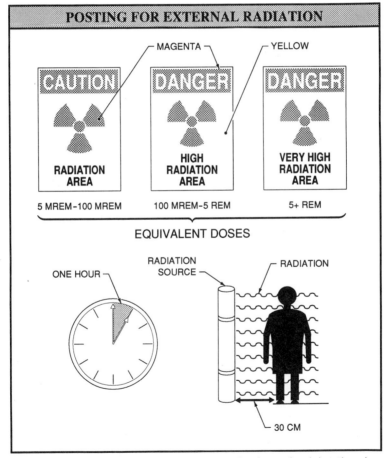

POSTING FOR EXTERNAL RADIATION

Figure 7-3. Posting for external radiation is determined by the dose equivalent received in one hour at 30 cm from the radiation source.

(Attachment 1) are present shall be clearly and conspicuously posted with a sign that identifies the radiological conditions which exist (e.g., "Airborne Radioactivity Area"). The type of sign used shall be consistent with the radiation protection control policies established at the facility and may be selected by the contractor with the approval of the field organization.

Posting for Surface Contamination. The access to any area where surface contamination levels greater than 10 times those specified in Attachment 2 are present shall be clearly and conspicuously posted with a sign that identifies the radiological conditions which exist (e.g., "Contamination Area"). The type of sign used shall be consistent with the radiation control policies established at the facility and may be selected by the contractor with the approval of the field organization.

Protective Clothing

Certain tasks in the nuclear power plant require protective clothing. Protective clothing is designed to minimize contamination and radiation exposure of personnel. Cloth, plastic, or other material is used for protective clothing depending on the plant and procedures. Protective clothing is designed for the protection of the body, head, hands, feet, and respiratory system of workers in radiological areas. For maximum protection, several layers of protective clothing may be required in areas of the nuclear power plant with high contamination. This requirement is based on the principle of ALARA. For example, in addition to a pair of protective coveralls for the body, outer coveralls with low levels of fixed contamination may be required to provide additional protection when working in areas of high contamination. Outer protective clothing with low levels of fixed contamination must remain in the designated areas and are marked for outer use only. Hard hats, hoods, eyeshields, and surgical caps provide protection for the head. Cotton liners and rubber gloves provide protection for the hands. Inner shoe covers, outer shoe covers, and rubber boots provide protection for the feet. Masks and hoods protect against airborne radioactivity for respiratory protection. Plastic raingear may also be required over protective clothing to protect against moisture when performing assigned tasks. Special protective clothing may be required for certain tasks in the nuclear power plant.

The type of protective clothing required for a particular task is determined by health physics personnel and specified on the RWP. Dosimetry required is also specified on the RWP. After use, protective clothing is deposited into marked containers for proper disposal. The proper sequence of donning and removing protective clothing is required to provide maximum protection of personnel from radioactive materials.

> **Warning:** Protective clothing requirements and donning and removing procedures will vary from plant to plant. Always follow specific plant procedures.

Donning Protective Clothing. Donning of protective clothing is performed in designated dressing areas in the nuclear power plant. The sequence and procedure for donning protective clothing must be followed.

Protective clothing worn by workers in nuclear power plants include surgical caps, respirators (if required), coveralls, glove liners, rubber gloves, inner and outer shoe covers, and hard hats. Hard hats are required according to plant procedures.

Tape is securely fastened to provide a tight seal on joining parts of protective clothing. Identification badges and keycards are worn inside the protective clothing. In some restricted areas, exceptions may be required as instructed by plant security. Dosimetry required is specified by health physics personnel.

Removing Protective Clothing. The removal of protective clothing must follow a prescribed sequence to prevent contamination. Each required step eliminates contact with potentially contaminated protective clothing. See Figure 7-4. Procedures for protective clothing removal may vary slightly from one nuclear power plant to another. The following general procedures are used when removing protective clothing:

1. Remove hard hat. Place in proper receptacle.

2. Remove ankle tape.

3. Remove closure tape.

4. Remove rubber glove tape.

5. Remove outer shoe covers.

6. Pull first glove off by fingertips. Insert finger in second and remove inside out.

7. Remove dosimetry.

Remove hard hat.

Place dosimetry inside hard hat.

Place hard hat top down on step-off pad.

Figure 7-4. Protective clothing must be removed in sequence.

8. Remove hood or surgical cap.

9. Remove coveralls.

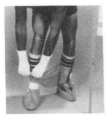 

10. Remove inner shoe cover. Step on step-off pad. Remove other inner shoe cover.

12. Frisk before stepping off step-off pad. Frisk dosimetry and hard hat.
13. Have health physics survey all materials exiting area.
14. Contact health physics if contamination is suspected.

11. Remove glove liners same way as rubber gloves.

Figure 7-4. Continued

1. Place contaminated yellow hard hat used in controlled areas in designated receptacle. (Check with health physics personnel.)

2. Remove ankle tape.

3. Remove closure tape.

4. Remove rubber glove tape.

5. Remove outer shoe covers.

6. Remove first rubber glove by pulling fingertips. Remove second rubber glove by inserting finger in glove and pulling the glove inside out.

7. Remove dosimetry. Remove white hard hat. Place dosimetry inside white hard hat. Place white hard hat on step-off pad with top down.

8. Remove hood or surgical cap.

9. Remove coveralls.

10. Remove inner shoe cover and step on step-off pad. Remove other inner shoe cover.

11. Remove glove liners using the same procedure as removing the rubber gloves.

12. Frisk before stepping off step-off pad or proceed to frisking area as required. Frisk dosimetry and hard hat as required.

13. All materials and tools exiting work area should be checked for contamination by health physics personnel.

14. Contact health physics if there is reason to suspect any contamination problems.

Personnel Monitoring Devices

Personnel monitoring devices required for specific tasks are specified on the RWP. Personnel monitoring devices commonly used are TLDs, film badges, direct-reading pocket dosimeters, and/or digital/alarming dosimeters. Personnel monitoring devices are required of all personnel entering controlled areas. Direct-reading pocket dosimeters are normally read after each shift. TLD and film badge readings are normally processed monthly, quarterly, or as required by the specific nuclear power plant. Digital/alarming dosimeters are primarily used in high radiation areas and/or for emergencies.

Specialized dosimetry may be specified by health physics on the RWP. For example, an electrician is assigned to rewire a crane motor. The motor is located behind a large vertical support. The radiation from the front of the vertical support is measured by the TLD or pocket dosimeter. However, in this case the back of the vertical support is highly radioactive. Therefore, the RWP specifies pocket dosimeters taped to the sleeves of the protective clothing and ring dosimeters to be worn on both hands. This measures radioactive exposure that would not have been measured with TLDs or film badges normally required. Other dosimetry may be required to measure exposure to extremities.

Loss or malfunction of dosimetry must be reported to health physics personnel immediately. If the reading on a direct-reading pocket dosimeter is off-scale, other personnel in the area must first be notified to check their readings. The malfunction should then be reported to health physics personnel. Health physics personnel will determine the amount of exposure received. A new direct-reading pocket dosimeter will be issued as required. Loss or malfunction of a TLD or film badge should also be reported immediately to health physics. Before the person can resume work in a controlled area, the dose must be determined by health physics. A new TLD will be issued. Permission to resume work is required from the appropriate health physics personnel. Loss of digital/alarming dosimeters are reported using the same procedures as direct-reading pocket dosimeters.

RADIATION WORK PERMITS

Radiation work permits (*RWPs*) are documents used to specify information regarding access, job description, special instructions, radiation protection requirements, ALARA evaluation, and approval of tasks in controlled areas of a nuclear power plant facility. The three types of RWPs are specific, area, and group. Radiation work permit instructions must be followed at all times in a nuclear power plant. Any special instructions are noted on the RWP. ALARA procedures are monitored using evaluation information provided on RWPs. See Figure 7-5.

WASHINGTON PUBLIC POWER SUPPLY SYSTEM

RADIATION WORK PERMIT

WNP- *2*

RWP NO. $\boxed{2}$-$\boxed{8}$$\boxed{8}$-$\boxed{0}$$\boxed{0}$$\boxed{0}$$\boxed{0}$$\boxed{1}$
TIME/DATE *2/3/88 0800*
TERMINATION DATE *UNKNOWN*
EXTENDED TO _____ BY _____

INITIATED BY *T. Jones* PHONE EXT. *8330* WORK GROUP *TRAINING*
JOB LOCATION BLDG./ELEV. *PSF/EOF* LOCATION *RM 120*
JOB DESCRIPTION *Training Exercise - obtain and don protective equipment as prescribed on this RWP: enter simulated posted area after reviewing RWP and survey map and making required entries on RWP Access Log Sheet and Radiation Exposure Card.*

EST MANHOURS				EST CREW SIZE				OTHER:
MECH	ELEC	I&C	OTHER	M	E	I&C	O	

RADIOLOGICAL CONDITIONS

RWP TYPE: SPECIFIC ☐   AREA ☐   GROUP ☑

RADIATION:
GENERAL AREA _*20*_ mrem/hr (TYPE) *B-8* LOCATION _____
CONTACT _*85*_ mrem/hr (TYPE) *B-8* LOCATION *PILLAR*
CONTAMINATION: *1K-6K* dpm/100cm² or _____ mrad/hr
AIRBORNE ACTIVITY *4.5 E-10* uci/cc INITIAL SURVEY NO. *PSF-0001*
SURVEY PERFORMED BY *T. Jones* TIME/DATE *2/3/88 0700*
COMMENTS: *SEE ATTACHED SURVEY MAP FOR CURRENT RADIOLOGICAL CONDITIONS AND SPECIFICS*

SPECIAL INSTRUCTIONS	RADIATION PROTECTION REQUIREMENTS	
STAYTIME: *UNLIMITED*	COVERAGE	__ CONTINUOUS __ PERIODIC ☑ INITIAL SURV. ONLY __ SEE SPECIAL INSTR. __ SURVEY FREQUENCY __
HP APPROVAL REQUIRED TO REMOVE MATERIAL FROM CONTAMINATION AREA		
WHOLE BODY FRISK REQUIRED UPON EXIT FROM CONTAMINATION AREA		
CONTACT HP PRIOR TO EACH ENTRY		
- *AVOID HOT SPOT LOCATION WHENEVER POSSIBLE*	MONITORING	☑ TLD ☑ 0-500 mR DOSI OR ☑ HIGH-RANGE DOSI __ SEE SPECIAL INSTR.
- *ZERO EITHER A 0-500 mr or 0-5R DOSIMETER*		
- *BRING THIS BOOK WITH YOU INTO THE SIMULATED POSTED AREA*		
- *CONTACT AN INSTRUCTOR IF YOU HAVE ANY QUESTIONS*	BODY	__ LAB COAT ☑ 1 PAIR COVERALLS __ 2 PAIR COVERALLS __ WATERPROOF __ SEE SPECIAL INSTR.
- *EXIT THE AREA UNDER OBSERVATION BY AN INSTRUCTOR*		
	HEAD	☑ HOOD __ WATERPROOF COVER __ EYE SHIELD __ CAP __ SEE SPECIAL INSTR.
ALARA EVALUATION	HAND	☑ COTTON LINERS __ COTTON WORK GLOVES ☑ RUBBER GLOVES _1_ PR. __ SEE SPECIAL INSTR.
ManRem ESTIMATE _____		
DATA ATTACHED ☐		
EVALUATION COMPLETED BY *T. Jones* DATE *2/3/88*	FEET	☑ RUBBER OVERSHOES __ RUBBER BOOTS ☑ PLASTIC BOOTIES __ SEE SPECIAL INSTR
APPROVAL		
JOB SUPERVISOR *R Smith* DATE *2/3/88*	RESPIRATORY	__ FULL FACE NP __ SUPPLIED AIR __ SCBA __ AIR HOOD __ SEE SPECIAL INSTR.
HP SUPERVISOR *J.A. Brown* DATE *2/3/88*		
TERMINATED BY _____ DATE _____		

WP062 R2 (12/85)   DIST: WHITE-RWP FOLDER, CANARY-POST @ ACCESS, PINK-JOB SUPERVISOR, GOLDENROD-SHIFT MANAGER
930076

Washington Public Power Supply System

Figure 7-5. Radiation work permits are required in controlled areas of nuclear power plants.

Information regarding contamination, radiation levels, and radioactivity of a work area is specified on a survey map and record. Each area is identified using abbreviations and symbols for level and location. This alerts the worker to potential hazards of performing tasks listed on the RWP.

Supplementary sheets are attached to RWPs if additional information or changes are required. Supplementary sheets can be used only with specific and area RWPs. Supplementary sheets cannot be used with group RWPs.

Specific RWPs

Specific RWPs are documents used to specify the performance of a particular task such as refueling, breeching radioactive systems, and dry well entry for maintenance. This RWP specifies conditions, termination date, and other requirements as it relates to the assigned task.

Area RWPs

Area RWPs are documents used to authorize various tasks in a specific area of the nuclear power plant such as 471 containment building. Areas in a nuclear power plant are indicated by elevation above sea level or the plant area. For example, the dry well in the containment building is located at 503'. See Figure 7-6. Grade level elevation of different nuclear power plants will vary depending on the geographic location of the plant. Grade level of a nuclear power plant located on a river in a valley is at a lower elevation than a nuclear power plant next to a lake on higher ground. The radiological conditions of an area specified on an area RWP are generally the same for all tasks.

Group RWPs

Group RWPs are documents that authorize a group of trades-workers or a particular craft for repeated tasks in areas of the nuclear power plant where there are consistent radiation levels. Group RWPs are used for tasks such as routine inspections and

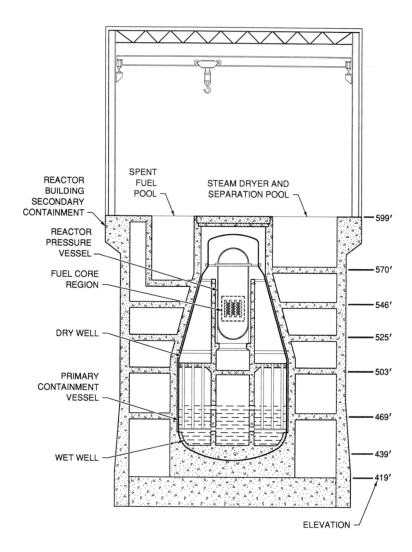

Figure 7-6. Areas in nuclear power plants are indicated by elevation numbers.

facility modification. For example, a group RWP is used for electricians adding lighting or extending communications lines in a specific area of the nuclear power plant.

EMERGENCY PROCEDURES

Emergency procedures are designed for each specific nuclear power plant. These procedures, using audio and visual alarms, alert personnel of conditions that require quick, decisive action. Different alarms are used in certain plant areas. Actions required are determined by the area and type of alarm. For example, the worker should leave the immediate area and call health physics when the amber light of an area radiation monitor (ARM) is activated. See Figure 7-7.

In addition to the emergency procedures designed for each specific nuclear power plant, the NRC also has established four emergency action levels that each nuclear power plant must follow during emergencies. *Emergency action levels* are specific levels of nuclear power plant emergencies and response requirements established by the NRC. These emergency action levels are notification, alert, site emergency, and general emergency. Unless controlled by the plant owner, lower emergency action levels can elevate into higher and much more serious levels.

Notification

The first emergency action level is the "notification of unusual event" which indicates a "potential degradation of the level of the safety of the plant." This level is not serious enough to cause a release of radioactivity "unless further degradation of safety systems occur." However, this level requires notification of appropriate state and local authorities. The notification level ends when the plant owner corrects the problem and states the situation is under control. If the emergency does not progress to the next level, a report as required is filed with appropriate authorities.

Alert

The second emergency action level is an "alert" for "an actual or potential substantial degradation of the level of safety of the

EMERGENCY PROCEDURES			
Alarm	**Indication**	**Location Alarm Can Be Heard**	**Action**
CAM (Continuous Air Monitor)	Audio - Bell Visual - Red Flashing light	Local Area	Leave immediate area. Call Health Physics.
ARM (Area Radiation Monitor)	Audio - None Visual - Amber	Local Area Control Room	Leave immediate area. Call Health Physics.
Criticality	Audio - Buzzer (Intermittent)	Local Area Control Room	Leave area covered by monitor. Report to Access Control.
Primary Containment Evacuation	Audio - Yelp Tone Visual - None	Local Area Control Room	Leave primary containment. Report to Health Physics personnel at access to containment.
Alert	Audio - Pulsed Tone (followed by announcement) Visual - None	All areas within the protected area Control Room	Listen for announcement. 1. Building evacuation 2. Plant evacuation 3. Adverse weather conditions 4. Fire 5. Etc.
Protected Area Controlled Evacuation	Audio - Alert Alarm (pulsed tone) Visual - None	All areas within the protected area Control Room	Report to Service Building lunchroom unless assigned to emergency response station. Individuals reporting to lunchroom provide nature and status of work being performed in plant prior to evacuation.
Protected Area Controlled Evacuation	Audio - Wailing Siren Visual - None	All areas within the protected area Control Room	Report to Plant Support Facility unless directed to other evacuation center.
Owner-Controlled Area	Audio - Two Alarms 1. Wailing Siren 2. Constant Tone Visual - None	All areas within the protected area Control Room Owner-controlled area and nearby facilities	Report to Plant Support Facility unless directed to other evacuation center. Report to Warehouse parking lot behind Supply System Headquarters unless directed to other evacuation center.
HNES: High noise area evacuation system	Audio - The existing alarm at greatly increased volume Visual - White Strobe	Local - high noise area	For: 1. Alert Tone - Call control room for instructions. 2. Evacuation Alarms - Follow prescribed action.

Figure 7-7. Emergency procedures are designed for each nuclear power plant.

plant." At this level, the release of radioactivity is "limited to small fractions of the EPA Action Guideline exposure levels." This level requires the plant owner to use plant personnel teams for monitoring the status of the plant. Appropriate authorities are informed of the plant status a minimum of every 15 minutes. Any release of radioactivity and the effects of weather conditions on the spread of radioactive material are also reported promptly. The alert level ends when the plant owner states the situation is under control. If the emergency does not progress to the next level, a report as required is filed with appropriate authorities.

Site Emergency

The third emergency action level is a "site emergency." This level occurs when "there are likely major failures of plant functions needed for protection of the public." The plant owner must start off-site monitoring for radiation. An emergency operations center is set up off-site near the plant. A technical support center is set up on-site. Plant officials provide information regarding the status of pressure levels in the reactor vessel to NRC and appropriate state and local authorities. Plant officials also provide information regarding projected amounts of radiation to be released and estimated doses based on plant and weather conditions. The site emergency level ends when the plant owner states the situation is under control. If the emergency does not progress to the next level, a report as required is filed with appropriate authorities.

General Emergency

The fourth emergency action level is a "general emergency." This is the most serious of emergency action levels. This level occurs when there is "actual or imminent substantial core degradation or melting with potential for loss of containment integrity." Action taken during the fourth emergency action level is similar to the action taken during a third emergency action level. In addition to reports of plant conditions to NRC and state and local authorities, a plant official briefs the news media. The emergency ends when the situation is brought under control by the required authorities at the emergency site. A report as required is filed with appropriate authorities.

Federal Radiological Emergency Response Plan

In the event of a nuclear power plant emergency or nuclear accident, specific government agencies are assigned to assist in a coordinated effort according to the Federal Radiological Emergency Response Plan (FRERP). Each of these government agencies has an area of expertise for which it is responsible. The FRERP was formed, in part, as a response to the Three Mile Island accident in Pennsylvania. The FRERP includes the following agencies:

- The Nuclear Regulatory Commission (NRC)
- The Federal Emergency Management Agency (FEMA)
- The Department of Defense (DOD)
- The Department of Energy (DOE)
- The Environmental Protection Agency (EPA)
- The Department of Transportation (DOT)
- The Department of Health and Human Services (HHS)
- The Department of Housing and Urban Development (HUD)
- The National Communications System (NCS)
- The Department of Agriculture (USDA)
- The Department of the Interior (DOI)
- The National Oceanic and Atmospheric Administration (NOAA)

PLANT PROCEDURES MANUAL

A plant procedures manual is required for every nuclear power plant as a part of the quality assurance program specified and required by the NRC. A plant procedures manual specifies and standardizes the required procedures to ensure the safe and efficient operation of the plant. The plant procedures manual is divided into separate volumes that include information pertaining to the following areas: security procedures, administrative procedures, operating procedures, abnormal condition procedures, testing procedures, fuel handling and refueling procedures, emergency procedures, maintenance, and health physics. Plant administration has all volumes of the plant procedures manual. Specific departments or supervisory personnel have volumes that apply directly to their area.

Chapter 8

Decommissioning and Waste Disposal

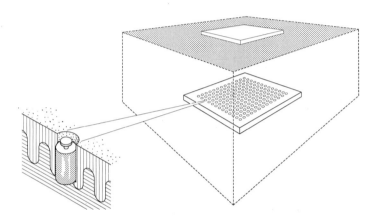

Nuclear power plants are the largest producers of radioactive waste in the United States. Radioactive waste is classified as either high- or low-level. Spent fuel is an example of high-level radioactive waste. Contaminated trash, gloves, and protective clothing are examples of low-level radioactive waste. All radioactive waste must be disposed of per the Nuclear Waste Policy Act of 1982.

When nuclear power plants are no longer efficient to operate, they are permanently shut down (decommissioned). The decommissioning process must follow the procedures established by the Nuclear Regulatory Commission (NRC). All fuel must be removed and properly disposed of and the nuclear reactor must be filled with concrete and sealed. Tradesworkers play a vital role in the work that must be done when decommissioning a nuclear power plant.

DECOMMISSIONING OF NUCLEAR FACILITIES

During the 1940s, a number of atomic reactors were built and operated to produce plutonium for atomic weapons programs. These plants, built with the technology of the 1940s, are not as practical today. For example, the B Reactor at the Hanford site near Richland, Washington, was constructed in 1943 and put into operation in the fall of 1944. It was shut down in 1968 and retired in 1979. See Figure 8-1. Additionally, some nuclear power plants are reaching the end of their projected life span. Utility com- panies can apply for extensions of plant operating licenses. However, many plants are now rapidly approaching obsolescence and must be decommissioned. *Decommissioning* is the process of permanently shutting down a nuclear facility to prevent the exposure of radiation to the public. The decommissioning process

U.S. Department of Energy, Richland Operations Office

Figure 8-1. Nuclear facilities must be decommissioned to provide maximum safety to the public.

must be in strict adherence to procedures established by the Nuclear Regulatory Commission (NRC). Decommissioning is a lengthy procedure.

Decontamination and Decommissioning

In the trades, the expression "D and D" is used to describe the process of decontamination and decommissioning of nuclear facilities. In this process, all fuel is removed from the reactor and sent to a reprocessing or nuclear waste disposal facility. This includes spent fuel, which accounts for approximately 99% of the radioactivity in a nuclear power plant. Fuel and plutonium processing plants are also decommissioned.

A large amount of cost and time is required to complete the decommissioning and nuclear waste disposal process. This effort requires the skills of thousands of electricians, plumbers, carpenters, sheet metal workers, and other tradesworkers. It has been estimated that as much as 150 billion dollars will be required to decommission and clean up the 16 nuclear facility sites slated for decommissioning. This amount exceeds the cost of the Apollo Space Program's lunar landing and approaches the expenditure required for the interstate highway system. Nuclear facilities are decommissioned by mothballing, dismantlement, and entombment.

Mothballing. *Mothballing* is the process of shutting down a nuclear facility and sealing off high radiation areas. All equipment is secured and the structure is left standing. The facilities are fenced and carefully monitored for radiation levels and possible intruders. Radiation in the sealed areas is reduced to safe levels over a 30- to 50-year period. The structure and equipment are then dismantled as required. After meeting federal, state, and local requirements, the land is reclaimed for other uses. There are several research and commercial nuclear facilities currently in mothball status.

Dismantlement. *Dismantlement* is the process of removing radioactive materials and dismantling nuclear facilities and equipment. In this process, the nuclear license is terminated and the ground is inspected for residual radioactivity. After meeting federal, state, and local requirements, the land is reclaimed for other uses.

Entombment. *Entombment* is the process of completely sealing off a nuclear reactor using concrete, steel, and other materials. The reactor is entombed for as long as the radiation poses a hazard, and to allow sufficient radioactive decay. Entombment was used after the Chernobyl nuclear accident to allow radiation levels to decrease and to protect further emission of radioactivity. The facilities are fenced and carefully monitored for radiation levels and possible intruders.

RADIOACTIVE WASTE

Radioactive waste consists of radioactive materials used in the fields of medicine, science, agriculture, and industry. The handling of these wastes varies depending on the quantity and level of radioactivity. All radioactive wastes must be handled, stored, and disposed of properly. The greatest amount of radioactive waste is produced by nuclear power plants. Radioactive waste is classified as high-level radioactive waste or low-level radioactive waste.

High-level Radioactive Waste

All nuclear power plants in the United States, including producers for civilian and military uses, have produced high-level radioactive waste material as a by-product of their operation. High-level radioactive waste includes spent (used) fuel rods as well as radioactive waste material from weapons processing facilities. Additionally, approximately one third of the fuel rods in reactor cores are replaced each year.

High-level radioactive waste is produced in small quantities compared to the total amount used in a fission reactor. For example, approximately 30 metric tons of spent fuel is produced from a 1000 MW nuclear power plant. Of this amount, approximately 2000 pounds are classified as high-level radioactive waste.

In the fission process, different isotopes are produced. Some isotopes such as plutonium can be reprocessed for reuse. However, many fission products and transuranic elements serve no use after the fission process. A *transuranic element* is an isotope created in the fission process that is heavier than uranium, with an atomic number above 92. Transuranic elements in spent fuel include isotopes of plutonium, neptunium, americium, and curium. Some radioisotopes have very short half-lives (seconds); others have very long half-lives (thousands of years). The treatment of high-level radioactive waste is very controversial. Long half-lives will make high-level radioactive waste an issue for many years to come.

In one year, a 1 million kW nuclear power plant uses approximately 100 tons of uranium fuel. With most nuclear power plants replacing one third of the fuel, approximately 33 tons of spent fuel must be removed and replaced.

Spent fuel rods change little physically during the fission process. However, radioisotopes change a great deal in the fission process. Most of the radioisotopes released from the ^{235}U leave ^{238}U. Even though the fuel is spent, heat and radiation are emitted for a long period of time after removal from the reactor core. To facilitate the cooling process, the spent fuel rods are placed underwater in a storage pool. Storage pools are located in or near the nuclear power plant. See Figure 8-2.

Storage pools are 40' deep and completely filled with water. The water cools spent fuel rods and absorbs neutrons. Storage pools are used as a temporary storage area to reduce the level of radioactivity in the spent fuel. Chemical inhibitors in the water are used to eliminate the possibility of a chain reaction occurring in the storage pools. Spent fuel rods are stored for approximately 10 years in storage pools. After that length of time, radioactivity has decreased by 90%.

Spent fuel rods are currently stored in storage pools at each nuclear reactor site. Some nuclear power plants are increasing

U.S. Council for Energy Awareness

Figure 8-2. Spent fuel rods are temporarily placed in storage pools. Temporary storage cannot exceed 25 years.

storage capacity by expanding present facilities, redesigning the racking systems to hold more rods, or sharing nearby nuclear power plant facilities to increase storage.

The U.S. Congress has passed legislation allowing temporary storage only of spent fuel rods. Temporary storage cannot exceed

25 years. Therefore, permanent disposal sites must be constructed to store long-lived, high-level radioactive waste.

Low-level Radioactive Waste

Low-level radioactive waste is generated by more than 20,000 government facilities, companies, and universities as a part of their routine activities. Examples of low-level radioactive waste include luminous watch dials, smoke alarms, radiopharmaceuticals, discarded manufacturing materials, used medical supplies, and contaminated paper, trash, gloves, and other protective clothing from nuclear power plants. Approximately one half of the low-level radioactive waste in the United States is generated by nuclear power plants. Most of the radioactive materials have a very short half-life, and radioactive decay occurs in months or years.

HIGH-LEVEL RADIOACTIVE WASTE DISPOSAL

Spent fuel bundles containing radioactive ashes from the fission process can be encapsulated in containers and disposed of as if they were high-level radioactive waste. A solution for the proper disposal of the amount of high-level radioactive waste accumulating across the country and proper protection from the potential hazards was needed. Studies were conducted to determine the best approach for disposal. The Nuclear Waste Policy Act of 1982 cited geologic repositories as the method of storing high-level radioactive waste. These repositories are deep underground. They are designed to prevent radioactive material from reaching the environment in solid form or from contact with water.

Although stable and small in volume, high-level radioactive waste is very radioactive and will remain a hazard for hundreds of years. Once solidified and buried in deep geologic repositories, the main risk is that groundwater could become contaminated and later be used for human consumption. Studies have shown that the risk from solidified high-level radioactive waste in a geologic repository after 100 years is much less than that from nonradioactive chemicals currently used by society such as barium, arsenic, chlorine, phosgene, and ammonia. See Figure 8-3.

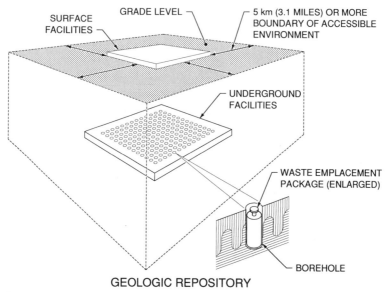

Figure 8-3. Solid high-level radioactive waste is stored underground in geologic repositories per the Nuclear Waste Policy Act of 1982.

High-level radioactive wastes must be placed in safe, carefully monitored federal repositories. Two sites have been selected: the Waste Isolation Pilot Plant in the salt caverns of New Mexico and the Nevada Test Site.

It is estimated that currently there are 45 million gallons of liquid radioactive waste contained in large underground tanks and over 200 billion gallons held in pits and ponds. Storage is also provided by tunnels dug deep underground in a rock media that will isolate the radioactive waste from any available groundwater for at least 10,000 years, long after the container has disintegrated.

Spent fuel is unloaded from the reactor vessel and transported to the storage pool by specialized remote handling equipment operating under 20′ of water. High-level radioactive waste destined for deep underground burial in Nevada is loaded into a primary container using remote manipulators in a shielded cell.

These containers are placed in a shielded cask that is then transported to the underground tunnel for emplacement. These cask/containers will be monitored periodically for leakage for approximately 50 years. Then, if the decision is made not to retrieve the high-level radioactive waste for some as yet unknown use, the tunnels will be backfilled and the site secured against intrusion by the public.

High-level liquid radioactive waste from reprocessing plants is currently being stored in tanks located in the Hanford Reservation in southeastern Washington and the Savannah River Plant in South Carolina. Leaks have been a continuing problem with this type of storage facility. All of the old tanks are being pumped out and the liquid radioactive waste transferred to double-shell tanks or to a salt cake for dry storage. This has solved the leakage problem. Since 1970, all liquid radioactive waste has been stored in double-shell tanks. See Figure 8-4.

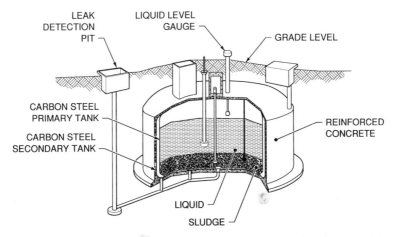

Figure 8-4. High-level liquid radioactive waste is stored underground in double-shell tanks.

Deep underground storage is limited to high-level, long-lived radioactive material. This material will still be radioactive thousands of years from now. The following three barriers to the migration of fission products are required:

- the initial container of the waste
- the shielded cask
- the rock media in which the containers are placed

The rock media selected for the storage site must be impervious to migration of the fission products after the first two barriers have disintegrated. Stringent requirements have been applied in the site selection process to assure the selection of a rock media that will contain the material for the thousands of years necessary for the long-lived radioactivity to die off.

Vitrification

Vitrification is a method of changing radioactive material into a glass-like material to stabilize liquid radioactive wastes. A vitrification plant at the Hanford Nuclear Reservation in the state of Washington is scheduled for operation in 1999.

In the vitrification process, glass-forming materials are mixed with liquid radioactive waste to form borosilicate glass. *Borosilicate glass* is a silica-based glass that has boron added to it. Liquid radioactive waste in the form of a slurry is mixed with glass-forming material. A *slurry* is a concentrated liquid waste that contains solids. The slurry is pumped into a glass furnace and heated to 2100°F. The molten glass is poured into a stainless steel cylinder. The molten glass cools and forms borosilicate glass. The cylinder is capped by welding and shipped for final disposal.

LOW-LEVEL RADIOACTIVE WASTE DISPOSAL

Low-level radioactive waste is placed in plastic bags at the source facility and then loaded into containers such as plastic and steel drums. The outside of the containers are inspected to verify that the exterior is not contaminated. The containers are then transported to the disposal site and placed in pits and trenches. See Figure 8-5. Low-level liquid radioactive waste from the processing facilities is piped directly into the underground storage tanks or transported by tank trucks to storage pools.

Figure 8-5. Low-level radioactive waste is placed in containers that are then stored in pits and trenches.

Pits and trenches on the surface are used for containers holding contact-handled waste and other waste. *Contact-handled waste* is low-level radioactive waste composed of contaminated material such as gloves, paper wipes, tools, etc. Contact-handled waste is received from schools, laboratories, hospitals, etc.

The dimensions of the trenches vary depending on soil and water conditions. The trenches are backfilled and covered with impermeable soil. Permanent markers are used to identify the amount and radiation type of waste stored.

Low-level radioactive waste storage facilities are located in Barnwell, South Carolina; Beatty, Nevada; and Hanford, Washington. In all cases, these burial sites are located many miles from the nearest populated area.

Low-level radioactive waste disposal in the future will be regulated by the Low-Level Radioactive Waste Policy Act of 1980 and subsequent amendments. This act requires each state to be responsible for its own low-level radioactive waste beginning January 1, 1993. The act also endorses the creation of regional disposal facilities for efficiency through regional compacts. In addition, after January 1, 1993, any state can refuse to accept low-level radioactive waste from states that are not members of the compact. This act affects all 50 states and the District of Columbia.

TRANSPORTATION OF RADIOACTIVE MATERIALS

Hazardous waste materials from industrial chemicals, pesticides, poisons, and toxic materials are shipped throughout the United States daily. The volume of hazardous wastes generated in the U.S. in a single year is more than 70,000 times larger than the volume of all spent fuel from commercial nuclear power plants in the last 24 years. The shipping of hazardous waste accounts for approximately one in 5000 of the total shipments made. Of the hazardous wastes shipped, approximately 2% are radioactive materials. Radioactive materials shipped include materials used in medicine, consumer products, and industrial by-products. Shipment of radioactive materials used in nuclear power plants accounts for approximately 0.025% of all hazardous waste materials shipped.

The two regulatory agencies that have primary responsibilities for transportation of radioactive materials are the Nuclear Regulatory Commission and the Department of Transportation. These agencies work with state and local agencies as required for maximum safety and security.

The regulation of the transportation of radioactive materials is under the authority of the Department of Transportation. This authority includes the following:

- training of personnel
- labeling of radite (hazardous) materials
- procedures for shipment of materials
- shipping papers
- loading/unloading procedures
- use of warning placards

DOT regulations require nuclear material shipments to follow direct interstate routes that bypass large population areas if possible. Routes to be taken are approved by the NRC. Routes designated for transportation of radioactive materials are selected in conjunction with affected municipalities. Drivers hauling radioactive materials must be notified of the contents of the load. In addition, drivers hauling radioactive materials must have special driver training certification. All vehicles transporting radioactive materials must display warning placards according to DOT regulations. Some state and local governments have established specific requirements of notification, routes, and equipment required to transport radioactive materials.

The regulations and procedures established by the NRC and DOT are directed toward reducing the potential for release of radiation from radioactive materials in the event of an accident or hostile assault. This requires as safe a package as possible for the transportation of radioactive materials. The packaging standards specified by the NRC and DOT are based on the degree of hazard posed by the radioactive material, quantity of radioactive material, and state of the radioactive material (liquid, solid, or gas).

Most radioactive materials transported by truck are in the form of low-level waste. Low-level waste is commonly shipped in containers that provide sufficient shielding to drivers and handlers. These containers must provide a tight seal to prevent spills or other common shipping hazards. An average nuclear power plant generates approximately 10 to 45 truckloads of low-level radioactive waste per year. As the potential hazard of the radio-activity is increased, the transport packaging must be designed to provide more protection and shielding. Higher level radioactive materials packaging must withstand a severe accident without the loss of shielding or the release of radiation. The shipping of highly radioactive spent fuel requires special shipping casks, demanding the greatest protection for transportation. See Figure 8-6.

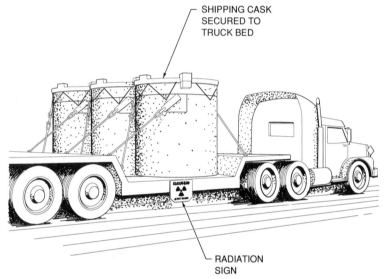

Figure 8-6. Highly radioactive spent fuel is transported in specially designed shipping casks to reduce the potential for accidents.

Shipping casks are designed to meet NRC requirements, which ensure the integrity of the cask in the event of an accident. Destructive tests are performed to check for failure of a cask including:

- complete immersion for eight hours
- exposure to 1475°F for eight hours
- 40" drop onto a 6" diameter pin
- drop from 30' onto a hard surface

These tests are designed to simulate the forces that may occur in an accident.

REPROCESSING

Reprocessing of spent fuel provides for the recovery of radioactive material for reuse. Reprocessing can recover up to 97% of the original uranium contained in spent fuel bundles. Reprocessing of spent fuel has been performed since the early days of nuclear experimentation and the Manhattan Project. Spent fuel rods, when removed from the reactor core, contain approximately 30% of the original fuel (uranium and plutonium). After removal from the reactor core, the spent fuel is cooled in storage pools. The spent fuel is removed as soon as possible to capture the maximum amount of usable chemical elements before radioactive decay.

Spent fuel is highly radioactive and requires special handling procedures. Most operations are performed with remote control equipment. After sufficient cooling, the spent fuel is shipped in sealed containers to storage facilities or reprocessing plants.

Spent fuel contains some usable nuclear material such as uranium and plutonium. Reprocessing involves chemical separation to recover usable uranium and plutonium. Fuel rods are cut in small lengths. Nitric acid is used during the dissolution process to separate fuel from reusable radioactive material and waste products of fission.

Radioactive material can be recovered many times. Reprocessed uranium is enriched similarly to new nuclear fuel in the enrichment plant. Reprocessed plutonium can be stored for use in advanced breeder reactors. Radioactive waste products are processed for storage in high-level radioactive waste storage facilities. See Figure 8-7.

The federal government prohibited the reprocessing of nuclear fuel in 1977. This was done as a security measure to prevent the possibility of reprocessed plutonium from being used to develop nuclear weapons. This ban was removed in 1981, but high costs have prevented the development of new commercial reprocessing plants.

Westinghouse Savannah River Company

Figure 8-7. Radioactive waste products are stored in high-level radioactive waste storage facilities.

Appendices

Appendix

CHEMICAL ELEMENTS

Name	Symbol	*Atomic Weight	Atomic Number	Name	Symbol	*Atomic Weight	Atomic Number
Actinium	Ac	[227]	89	Neon	Ne	20.183	10
Aluminum	Al	26.9815	13	Neptunium	Np	[237]	93
Americium	Am	[243]	95	Nickel	Ni	58.71	28
Antimony	Sb	121.75	51	Niobium	Nb	92.906	41
Argon	Ar	39.948	18	Nitrogen	N	14.0067	7
Arsenic	As	74.9216	33	Nobelium	No	[255]	102
Astatine	At	[210]	85	Osmium	Os	190.2	76
Barium	Ba	137.34	56	Oxygen	O	15.9994	8
Berkelium	Bk	[247]	97	Palladium	Pd	106.4	46
Beryllium	Be	9.0122	4	Phosphorus	P	30.9738	15
Bismuth	Bi	208.980	83	Platinum	Pt	195.09	78
Boron	B	10.811	5	Plutonium	Pu	[244]	94
Bromine	Br	79.909	35	Polonium	Po	[210]	84
Cadmium	Cd	112.40	48	Potassium	K	39.102	19
Calcium	Ca	40.08	20	Praseodymium	Pr	140.907	59
Californium	Cf	[251]	98	Promethium	Pm	[145]	61
Carbon	C	12.01115	6	Protactinium	Pa	[231]	91
Cerium	Ce	140.12	58	Radium	Ra	[226]	88
Cesium	Cs	132.905	55	Radon	Rn	[222]	86
Chlorine	Cl	35.453	17	Rhenium	Re	186.2	75
Chromium	Cr	51.996	24	Rhodium	Rh	102.905	45
Cobalt	Co	58.9332	27	Rubidium	Rb	85.47	37
Copper	Cu	63.54	29	Ruthenium	Ru	101.07	44
Curium	Cm	[247]	96	Samarium	Sm	150.35	62
Dysprosium	Dy	162.50	66	Scandium	Sc	44.956	21
Einsteinium	Es	[254]	99	Selenium	Se	78.96	34
Erbium	Er	167.26	68	Silicon	Si	28.086	14
Europium	Eu	151.96	63	Silver	Ag	107.870	47
Fermium	Fm	[257]	100	Sodium	Na	22.9898	11
Fluorine	F	18.9984	9	Strontium	Sr	87.62	38
Francium	Fr	[223]	87	Sulfur	S	32.064	16
Gadolinium	Gd	157.25	64	Tantalum	Ta	180.948	73
Gallium	Ga	69.72	31	Technetium	Tc	[97]	43
Germanium	Ge	72.59	32	Tellurium	Te	127.60	52
Gold	Au	196.967	79	Terbium	Tb	158.924	65
Hafnium	Hf	178.49	72	Thallium	Tl	204.37	81
Helium	He	4.0026	2	Thorium	Th	232.038	90
Holmium	Ho	164.930	67	Thulium	Tm	168.934	69
Hydrogen	H	1.00797	1	Tin	Sn	118.69	50
Indium	In	114.82	49	Titanium	Ti	47.90	22
Iodine	I	126.9044	53	Tungsten	W	183.85	74
Iridium	Ir	192.2	77	Unnilennium	Une	[266]	109
Iron	Fe	55.847	26	Unnilhexium	Unh	[263]	106
Krypton	Kr	83.80	36	Unniloctium	Uno	[265]	108
Lanthanum	La	138.91	57	Unnilpentium	Unp	[262]	105
Lawrencium	Lr	[256]	103	Unnilquadium	Unq	[261]	104
Lead	Pb	207.19	82	Unnilseptium	Uns	[262]	107
Lithium	Li	6.939	3	Uranium	U	238.03	92
Lutetium	Lu	174.97	71	Vanadium	V	50.942	23
Magnesium	Mg	24.312	12	Xenon	Xe	131.30	54
Manganese	Mn	54.9380	25	Ytterbium	Yb	173.04	70
Mendelevium	Md	[258]	101	Yttrium	Y	88.905	39
Mercury	Hg	200.59	80	Zinc	Zn	65.37	30
Molybdenum	Mo	95.94	42	Zirconium	Zr	91.22	40
Neodymium	Nd	144.24	60				

*A number in brackets indicates the mass number of the most stable isotope

NUCLEAR POWER ACRONYMS

AEC	Atomic Energy Commission
ALARA	as low as reasonably achievable
ANSI	American National Standards Institute
BWR	boiling water reactor
CANDU	Canadian deuterium-uranium
CFR	Code of Federal Regulations
DOD	Department of Defense
DOE	Department of Energy
DOI	Department of Interior
DOT	Department of Transportation
DSA	digital subtraction angiography
EPA	Environmental Protection Agency
ERDA	Energy Research and Development Administration
FEMA	Federal Emergency Management Agency
FRERP	Federal Radiological Emergency Response Plan
HHS	Department of Health and Human Services
HTGR	high temperature gas-cooled reactor
HUD	Department of Housing and Urban Development
LD	lethal dose
LMFBR	liquid metal fast breeder reactor
LWR	light water reactor
MRI	magnetic resonance imaging
NASA	National Aeronautics and Space Administration
NCS	National Communications System
NOAA	National Oceanic and Atmospheric Administration
NRC	Nuclear Regulatory Commission
PWR	pressurized water reactor
RWP	radiation work permit
SONO	sonography
SPECT	single photon emission computed tomography
TLD	thermoluminescent dosimeter
USDA	United States Department of Agriculture

DECIMAL AND MILLIMETER EQUIVALENTS

Fraction	Decimal	Millimeter	Fraction	Decimal	Millimeter
1/64	.015625	.397	33/64	.515625	13.097
1/32	.03125	.794	17/32	.53125	13.494
3/64	.046875	1.191	35/64	.546875	13.891
1/16	.0625	1.588	9/16	.5625	14.288
5/64	.078125	1.984	37/64	.578125	14.684
3/32	.09375	2.381	19/32	.59375	15.081
7/64	.109375	2.778	39/64	.609375	15.478
1/8	.125	3.175	5/8	.625	15.875
9/64	.140625	3.572	41/64	.640625	16.272
5/32	.15625	3.969	21/32	.65625	16.669
11/64	.171875	4.366	43/64	.671875	17.066
3/16	.1875	4.762	11/16	.6875	17.462
13/64	.203125	5.159	45/64	.703125	17.859
7/32	.21875	5.556	23/32	.71875	18.256
15/64	.234375	5.953	47/64	.734375	18.653
1/4	.25	6.350	3/4	.75	19.050
17/64	.265625	6.747	49/64	.765625	19.447
9/32	.28125	7.144	25/32	.78125	19.844
19/64	.296875	7.541	51/64	.796875	20.241
5/16	.3125	7.938	13/16	.8125	20.638
21/64	.328125	8.334	53/64	.828125	21.034
11/32	.34375	8.731	27/32	.84375	21.431
23/64	.359375	9.128	55/64	.859375	21.828
3/8	.375	9.525	7/8	.875	22.225
25/64	.390625	9.922	57/64	.890625	22.622
13/32	.40625	10.319	29/32	.90625	23.019
27/64	.421875	10.716	59/64	.921875	23.416
7/16	.4375	11.112	15/16	.9375	23.812
29/64	.453125	11.509	61/64	.953125	24.209
15/32	.46875	11.906	31/32	.96875	24.606
31/64	.484375	12.303	63/64	.984375	25.003
1/2	.5	12.700	1	1.0	25.400

U.S. CONVERSION FACTORS
LENGTHS AND AREAS

To Convert	Multiply By	Example
Lengths Inches to feet	.0833	36″ × .0833 = 2.9988 = 3′
Inches to yards	.0278	92″ × .0278 = 2.5576 = 2.56 yd
Feet to inches	12	5′ × 12 = 60″
Feet to yards	.33	12′ × .33 = 3.96 = 4 yd
Feet to miles	.000189	10,560′ × .000189 = 1.99584 = 2 mi
Areas Square feet to square inches	144	1.25 sq ft × 144 = 180 sq in.
Square feet to square yards	.111	18 sq ft × .111 = 1.998 = 2 sq yd
Square feet to acres	.00002296	1,742,400 sq ft × .00002296 = 40.005504 = 40 acres
Square yards to square feet	9	3 sq yd × 9 = 27 sq ft
Square yards to acres	.0002066	774,400 sq yd × .0002066 = 159.99 = 160 acres
Square miles to square feet	27,878,400	.01 sq mi × 27,878,400 = 278,784 sq ft
Square miles to square yards	3,097,600	.16 sq mi × 3,097,600 = 495,616 sq yd
Square miles to acres	640	.25 sq mi × 640 = 160 acres

Note: Various values are used for examples

U.S.–METRIC CONVERSION FACTORS LENGTHS AND AREAS		
To Convert	**Multiply By**	**Example**
Lengths Inches to centimeters	2.54	6″ × 2.54 = 15.24 = 15 cm
Inches to meters	.0254	945″ × .0254 = 24.003 = 24 m
Inches to millimeters	25.4	15″ × 25.4 = 381 mm
Feet to centimeters	30.48	15′ × 30.48 = 457.2 = 457 cm
Feet to kilometers	.0003048	5280′ × .0003048 = 1.609344 = 1.6 km
Feet to meters	.3048	21′ × .3048 = 6.4008 = 6.4 m
Feet to millimeters	304.8	1.5′ × 304.8 = 457.2 = 457 mm
Areas Square feet to square meters	.0929	27 sq ft × .0929 = 2.5083 = 2.5 sq m
Square yards to square meters	.8361	6 sq yd × .8361 = 5.0166 = 5 sq m
Square miles to square meters	2,589,600	.05 sq mi × 2,589,600 = 129,480 sq m
Square miles to square kilometers	2.590	25 sq mi × 2.590 = 64.75 sq km
Note: Various values are used for examples. To convert metric to U.S., reverse the process. For example, 381 mm ÷ 25.4 = 15″.		

U.S. SHORT TONS–METRIC TONS CONVERSIONS

Short Tons	Metric Tons	Short Tons	Metric Tons	Short Tons	Metric Tons	Short Tons	Metric Tons
1	0.907	26	23.587	51	46.266	76	68.946
2	1.814	27	24.494	52	47.174	77	69.853
3	2.722	28	25.401	53	48.081	78	70.760
4	3.629	29	26.308	54	48.988	79	71.668
5	4.536	30	27.216	55	49.895	80	72.575
6	5.443	31	28.123	56	50.802	81	73.482
7	6.350	32	29.030	57	51.710	82	74.389
8	7.257	33	29.937	58	52.617	83	75.296
9	8.165	34	30.844	59	53.524	84	76.204
10	9.072	35	31.751	60	54.431	85	77.111
11	9.979	36	32.659	61	55.338	86	78.018
12	10.886	37	33.566	62	56.245	87	78.925
13	11.793	38	34.473	63	57.153	88	79.832
14	12.701	39	35.380	64	58.060	89	80.739
15	13.608	40	36.287	65	58.967	90	81.647
16	14.515	41	37.195	66	59.874	91	82.554
17	15.422	42	38.102	67	60.781	92	83.461
18	16.329	43	39.009	68	61.689	93	84.368
19	17.237	44	39.916	69	62.596	94	85.275
20	18.144	45	40.823	70	63.503	95	86.183
21	19.051	46	41.731	71	64.410	96	87.090
22	19.958	47	42.638	72	65.317	97	87.997
23	20.865	48	43.545	73	66.225	98	88.904
24	21.772	49	44.452	74	67.132	99	89.811
25	22.680	50	45.359	75	68.039	100	90.718

Note: 1 U.S. short ton = 2000 lb
1 metric ton = 1 U.S. short ton × .907 185
1 metric ton = 1814.37 lb (2000 × .907 185 = 1814.37)

METRIC PREFIXES				
Prefix	Symbol	Power of 10	Unit	Example
pico	p	10^{-12}	$\dfrac{1}{1,000,000,000,000}$	1 picofarad = 10^{-12} farad
nano	n	10^{-9}	$\dfrac{1}{1,000,000,000}$	1 nanosecond = 10^{-9} second
micro	μ	10^{-6}	$\dfrac{1}{1,000,000}$	1 microcurie = 10^{-6} curie
milli	m	10^{-3}	$\dfrac{1}{1000}$	1 millirem = 10^{-3} rem
centi	c	10^{-2}	$\dfrac{1}{100}$	1 centimeter = 10^{-2} meter
deci	d	10^{-1}	$\dfrac{1}{10}$	1 decigram = 10^{-1} gram
kilo	k	10^{3}	1000	1 kilowatt = 10^{3} watts
mega	M	10^{6}	1,000,000	1 megaohm = 10^{6} ohms
giga	G	10^{9}	1,000,000,000	1 gigaelectronvolt = 10^{9} electronvolts

BIOLOGICAL EFFECTS OF ACUTE RADIATION DOSES

Dose Rems	Symptoms/ Effects	Latent Period	Characteristic Signs	Therapy	Convalescent Period	Incidence of Death	Time of Death	Cause of Death
0–25	no detectable effects (95–100%) minor blood changes (0–5%)	none 48 hours	none low white cell count	none none	none 1–2 weeks	0%		none
25–50	detectable changes in blood (95–100%)	24 hours	low white cell count low red cell count	blood transfusion	2–4 weeks	0%		none
50–100	detectable changes in blood (100%) nausea (0–5%)	12 hours 12 hours	low red cell count low white cell count vomiting	blood transfusion stomach relaxers	4–8 weeks 2–10 days	0%		none
100–200	nausea (5–50%) loss of body hair (0–10%)	3 hours 5–10 days	vomiting minor balding	stomach relaxers none	several weeks 6–12 months	0–10%	2–20 years	cancer
200–600	anemia nausea loss of body hair (10–50%) heart palpitations	1–12 days 2 hours 2–5 days 1 hour– 12 months	leukopenia purpura hemorrhage severe vomiting balding rapid pulse	blood transfusion antibiotics stomach relaxers reassurance	1–12 months	10–80%	1–2 months	hemorrhage infection heart failure cancer/leukemia
600–1000	severe anemia nausea-diarrhea severe loss of body hair heart palpitations/heart attack	1 hour 1 hour 2–3 days 1 hour	leukopenia purpura hemorrhage intestinal discharge severe balding rapid pulse heart stoppage	blood transfusion blood coagulants bone marrow transplants intravenous feeding adrenalin	extremely long	80–100%	2 days– 2 months	infection hemorrhage heart failure kidney failure respiratory failure cancer/leukemia
1000–100,000	pernicious anemia nervous system failure nausea-diarrhea heart, respiratory, excretory failure	1 hour 30 minutes 30 minutes 1 hour– 30 minutes	leukopenia purpura hemorrhage convulsions tremor-lethargy	blood transfusion bone marrow transplants sedatives	extremely long	99–100%	1 hour– 2 weeks	infection, hemorrhage heart failure kidney failure respiratory failure coma brain failure
100,000	complete malfunction	none	cell wall rupture	none	none	100%	instantly	molecular death

INFORMATION SOURCES

American Gas Association
1515 Wilson Boulevard
Arlington, VA 22209
(703) 841-8400

American Nuclear Society
555 N. Kensington
La Grange Park, IL 60525
(708) 352-6611

American Petroleum Institute
1220 L Street, N.W.
Washington, DC 20005
(202) 682-8000

Council on Environmental Quality
722 Jackson Place, N.W.
Washington, DC 20503
(202) 395-5750

Edison Electric Institute
1111 19th Street, N.W.
Washington, DC 20036
(202) 778-6400

Electric Power Research Institute
3412 Hillview Avenue
P.O. Box 10412
Palo Alto, CA 94303
(415) 855-2000

1019 19th Street, N.W.
10th Floor
Washington, DC 20036
(202) 872-9222

National Coal Association
1130 17th Street, N.W.
Washington, DC 20036
(202) 463-2625

National American Electric Reliability Council
101 College Road East
Princeton, NJ 08540-6601
(609) 452-8060

Nuclear Regulatory Commission
Office of Public Affairs
Washington, DC 20555
(301) 492-0240

U.S. Council for Energy Awareness
Suite 400
1776 I Street, N.W.
Washington, DC 20006-2495
(202) 293-0770
after-hours (202) 466-0246

Suite 644
60 East 42nd Street
New York, NY 10165-0198
(212) 599-1881
after-hours (212) 348-6622

U.S. Environmental Protection Agency
Office of Radiation Programs
401 M Street, S.W.
108 Northeast Mall
Washington, DC 20460
(202) 475-9600

U.S. Department of Energy
Assistant Secretary for Nuclear Energy
1000 Independence Avenue, S.W.
Room 5A-115
Washington, DC 20585
(202) 586-6450

Energy Information Administration
Director of Nuclear and Alternate
    Fuels
1000 Independence Avenue, S.W.
Room BG-057
Washington, DC 20585
(202) 586-8383

Energy Information Administration
National Energy Information Center
1000 Independence Avenue, S.W.
Room 1F-048–EI-231
Washington, DC 20585
(202) 586-8800

Glossary

accelerator: Device used to smash atoms by shooting a stream of high-speed particles at large atoms such as uranium.

acute exposure: Large doses of radiation received in a short time.

air monitor: Radiation detection device that continuously monitors air for airborne radioactivity at specific locations.

alpha particle: Positive-charged nuclear particle identical to a helium atom nucleus, which has two protons and two neutrons. An alpha particle is ejected at high speed in certain radioactive transformations.

Eberline Instruments
air monitor

alpha radiation: Radiation that is the most energetic (densely ionizing) but the least penetrating type.

antineutrino: Antiparticle that has no mass or electrical charge.

antiparticle: Particle with the same mass as another particle but opposite electrically and magnetically.

archaeological: Related to the scientific study of fossils, relics, and remains of past cultures.

area monitor: Radiation detection device that continuously monitors gamma radiation in specific locations. Area monitors are wall-mounted and contain audible and light alarms.

Dosimeter Corporation
area monitor

area RWP: Radioactive work permit used to authorize various tasks in a specific area of a nuclear power plant.

atom: Smallest building block of matter that cannot be divided into smaller units without changing its basic character.

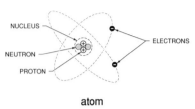

atom

atomic number: The number of protons in the nucleus of a chemical element.

atomic weight: The weight of an atom relative to a standard of carbon (^{12}C), which has an atomic weight of 12. The sum of the protons and neutrons in a chemical element equals the atomic weight.

audible: Capable of being heard.

B

background radiation: Radiation present in the Earth's atmosphere.

beta radiation: Radiation consisting of high-speed electrons emitted from the unstable nuclei of certain radioactive chemical elements.

binding energy: Energy required to overcome an atom's nuclear forces to release protons and neutrons.

biological: Related to life and living processes.

blackout: Loss of electrical power caused by an interruption of service from the supplier.

boiling water reactor (BWR): Type of light water reactor that operates with cooling water passing through the reactor core.

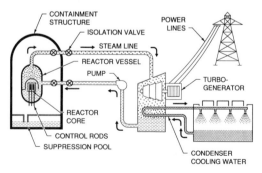

boiling water reactor (BWR)

borosilicate glass: Silica-based glass that has boron added to it.

breeder reactor: Type of nuclear reactor that uses liquid sodium, rather than water, for heat transfer. Breeder reactors produce fuel material.

brownout: Reduction of voltage from the supplier as a measure to prevent a blackout.

C

chain reaction: Process in which released neutrons from an atom strike and split other atoms, which repeat the procedure. Fission is maintained in a nuclear chain reaction.

chemical compound: Chemical elements containing atoms of two or more different types.

chemical element: Basic substances consisting of atoms of one type that, alone or combined with other chemical elements, constitute all matter.

chemical mixture: Mixture of chemical elements and/or chemical compounds.

chronic exposure: Small doses of radiation received over a long period of time.

contact-handled waste: Low-level radioactive waste composed of contaminated material such as gloves, paper wipes, tools, etc.

contamination: Presence of removable radioactive material in any place where it is not desired.

controlled area: Nuclear power plant area in which radioactive materials or radiation may exist.

control rod: Nuclear reactor part used to control the rate of the chain reaction in a reactor.

coolant: Liquid that absorbs and transports heat produced in a nuclear reactor.

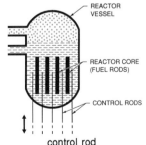

control rod

cooling water: Liquid used to remove heat from nuclear reactors and steam used in turbo-generators.

core: Central part of the nuclear reactor that contains the fuel elements.

critical mass: Smallest mass of fissionable material that will support a self-sustaining chain reaction under certain conditions.

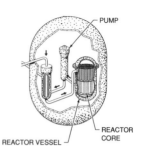

core

critical reaction: A nuclear reaction in which the average number of fission-producing neutrons is one.

curie: Unit used to measure the rate at which particles or rays are released (disintegrated) from a radioactive isotope. A curie equals 3.7×10^{10} disintegrations per second.

D

date analysis: Process of measuring the age of archaeological finds.

decommissioning: Process of permanently shutting down a nuclear facility.

decontamination: Removal of radioactive material from an undesired place.

delayed effect: Biological effect of radiation appearing months or years after exposure.

depleted: Lessened in quantity, content, power, or value.

digital: Numerical readout.

digital/alarming dosimeter: Radiation detection device that provides two ways of indicating radiation levels, digital readout or audible alarm.

direct-reading pocket dosimeter: Radiation detection device with a gas-filled ion chamber that constantly monitors gamma radiation and X-ray levels.

Dosimeter Corporation
digital/alarming dosimeter

disintegration: The act of breaking into parts.

dismantlement: Process of removing radioactive materials and dismantling nuclear facilities and equipment.

distribution substation: Electrical facility that provides switching and decreases voltage of electricity for end users.

dosimeter: Any device that measures doses of radiation.

dosimetry: All equipment used to detect and measure radiation emitted and doses received.

E

electromagnetic: The force or field created by passing an electric current through the core of a magnet.

electron: Negative-charged particles of matter that orbit around the nucleus of an atom.

electrostatic: Related to stationary charges of electricity.

emergency action levels: Specific levels of nuclear power plant emergencies and response requirements established by the Nuclear Regulatory Commission (NRC).

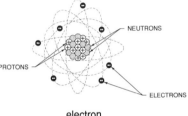

electron

entombment: Process of completely sealing off a nuclear reactor using concrete, steel, and other materials.

exponent: Number indicating a power of 10 in scientific notation.

external exposure: Biological effects caused by radiation exposure to external parts of the body.

F

film badge: Radiation detection device in the form of a badge that contains film sensitive to radiation. The film is exposed in proportion to the amount of radiation received.

Tech/Ops Landauer, Inc.

film badge

fission track dating: Date analysis method that records fission tracks on a film placed next to the material being dated. Fission tracks are counted to determine the amount of uranium 238 (^{238}U) or potassium 40 (^{40}K) left in the material.

Victoreen, Inc.

frisker

frisker: Radiation detection device used to detect and measure radiation when leaving a controlled area.

G

gamma radiation: High-energy electromagnetic energy waves that have more penetrating power than alpha radiation or beta radiation.

Geiger (Geiger-Muller or G-M) counter: Radiation detection device that detects the presence and measures the intensity of radiation through the ionizing effect of an enclosed gas.

Dosimeter Corporation

Geiger counter

genetic effect: Biological effect caused by radiation transferred from parent to offspring.

group RWP: Radiation work permit used to authorize a group of tradesworkers to perform repeated tasks in specific areas of a nuclear power plant.

H

half-life: Time required for a quantity of radioactive material to lose one-half its radioactivity. During this process, half of the atoms decay into another chemical element or isotope.

half-value layer: Thickness of shielding material that will stop half of the radiation from penetrating through.

hand and shoe monitor: Radiation detection device used to monitor radiation levels of hands and shoes.

high radiation level: Nuclear power plant area that has a dose greater than 100 mrem/hour but less than 5 rem/hour at 30 cm from the radiation source.

I

internal exposure: Biological effects caused by the ingestion of radioactive materials.

ion: An atom with a positive or negative electrical charge.

ionization chamber: Device that uses ionization of gases to measure radiation.

irradiation: Exposure to radiation such as X rays or a stream of neutrons.

isotope: Form of the chemical element.

L

latent period: Time between radiation exposure and its effect.

lethal dose (LD): Term used to describe the deadly effects of radiation exposure.

light water reactor (LWR): Nuclear reactor type that uses a liquid coolant (water or light water) that is pumped into the reactor vessel through the reactor core to remove heat.

M

mass number: Total number of protons and neutrons in the nuclei of an atom. It is used to identify a specific isotope.

milli: Metric prefix for one thousandth.

moderator: Nuclear reactor part that slows down neutrons in the fission process.

molecule: Two or more atoms joined together by forces.

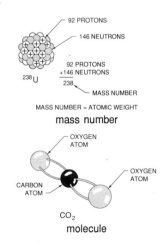

92 PROTONS

146 NEUTRONS

92 PROTONS
+146 NEUTRONS

238 U   238

MASS NUMBER

MASS NUMBER = ATOMIC WEIGHT

mass number

OXYGEN ATOM

OXYGEN ATOM

CARBON ATOM

CO_2

molecule

mothballing: Process of temporarily shutting down a nuclear facility and sealing off high radiation areas.

mutant: Significant and basic physical change to the chromosomes or genes of a plant or animal.

N

neutrino: An uncharged (neutral) particle with high penetrating power that is produced in a nuclear reaction.

neutron: Uncharged (neutral) particle contained in the nucleus of an atom.

neutron activation analysis: Detection process that can determine trace elements as small as one billionth of a part.

neutron radiation: Most highly penetrating and difficult to shield of all radiation types.

nuclear energy: Energy released from an atomic reaction.

nuclear fission: Energy released when an atom is split into two atoms.

nuclear forces: Forces that hold the nucleus of an atom together.

nuclear fusion: Energy released when two atoms are forced together to make a third atom.

nucleus: Heavy, dense center of an atom, which contains protons and neutrons and has a positive electrical charge.

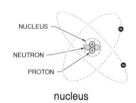

nucleus

O

overhead transformer: Transformer mounted on a pole to provide overhead electrical service to buildings.

P

pad-mounted transformer: Transformer placed on a concrete pad to provide lateral (underground) service to buildings.

photon: Particle of electromagnetic radiation that travels at the speed of light.

point source: A concentration of radioactivity in a small volume.

portable nuclear power plant: Nuclear power plant designed for usage in any location.

portal monitor: Radiation detection device that surrounds the body. G-M detectors are located in the frame.

positron: Particle that has the same mass as an electron but contains a positive electrical charge.

posted area: Nuclear power plant area in which radiation or contamination exceeds normal levels specified.

pressurized water reactor (PWR): Type of light water reactor that uses a reactor core surrounded by a steel "core barrel" or "shroud" reactor vessel.

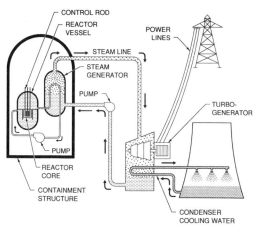

pressurized water reactor (PWR)

prime mover: Mechanical device used to drive generators.

prompt effect: Biological effect of radiation appearing shortly after exposure.

proton: Particle with a positive electrical charge contained in the nucleus of an atom.

Q

quality factors: Standards determined by the amount of biological damage a specific radiation can produce.

R

rad (radiation absorbed dose): Unit of absorbed dose equal to 100 ergs of energy per gram. It is a measurement of the ionizing radiation absorbed per gram of material.

radiation: Emission of high-energy particles (rays) from atoms with unstable nuclei.

radiation area: Nuclear power plant area that has a dose rate greater than 5 mrem/hour, but less than 100 mrem/hour at 30 cm from the radiation source.

radiation sickness: Any illness precipitated by exposure to radiation.

radiation work permit (RWP): Document used to specify information regarding access, job description, special instructions, radiation protection requirements, ALARA evaluation, and approval of tasks in controlled areas of a nuclear power plant facility.

radioactive decay: Loss of radioactivity by radioisotopes over a period of time measured in half-lives. During this process, the nucleus changes in stability into the form of another isotope or chemical element.

radioactivity: Process of spontaneous disintegration of radioactive nuclei.

radiocarbon dating: Date analysis method that measures radioactive carbon 14 (^{14}C) to determine the age of archaeological finds.

radioisotope: Atom that emits radiation.

radiotracer: Radioisotope used for tracing specific chemical elements.

reactor vessel: Nuclear reactor part made of heavily constructed steel that contains the entire reactor core.

reflector: Layer of material surrounding a nuclear reactor core, which reflects neutrons that would otherwise escape back into the core.

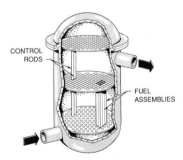

reactor vessel

rem (roentgen equivalent in man): Measurement of the amount of biological damage from a dose of radiation. It is equivalent in biological damage to 1 rad of 250 kW of X rays.

roentgen: Quantity of X rays or gamma radiation that will produce one electrostatic unit of positive or negative electrical charge in a cubic centimeter of dry air.

S

scanner: Detector that shows radiation introduced into the body.

scientific notation: Process of using powers of 10 to simplify math for very large or small numbers.

scintillation detector: Radiation detector device that senses and measures radiation striking certain chemicals causing flashes of light (scintillations), which are converted to digital impulses.

Ludlum Measurements, Inc.
scintillation detector

shielding: Any material that contains and absorbs radiation produced in the fission process in a nuclear reactor.

short ton: Measurement equivalent to 0.907 185 metric tons (t), or 2000 pounds.

slurry: Concentrated liquid waste that contains solids.

solid-state radiation detector: Radiation detection device that measures electrons released by radiation using semiconductors.

somatic effect: Biological effect from radiation that causes direct damage to cell molecules in the body.

specific RWP: Radioactive work permit used to specify the performance of a particular task.

stationary nuclear power plant: Nuclear power plant constructed in a permanent location.

step-down transformer: Transformer that decreases voltage of electricity for short distribution and end use.

step-up transformer: Transformer that increases voltage of electricity for long-distance transmission.

subcritical reaction: Nuclear reaction in which the average number of fission-producing neutrons is less than one.

supercritical reaction: Nuclear reaction in which the average number of fission-producing neutrons is more than one.

T

tails: Depleted material that is unusable in a nuclear reactor.

thermoluminescent dosimeter (TLD): Radiation detection device that measures X rays and beta, gamma, and neutron radiation using lithium fluoride crystals.

Tech/Ops Landauer, Inc.
thermoluminescent dosimeter

transmission substation: Electrical facility that provides switching and decreases high voltage for transmission of electricity to substations and heavy industry.

transuranic element: Isotope created in the fission process that is heavier than uranium, with an atomic number above 92.

turbo-generator: A steam turbine generator that processes steam used to produce electricity.

U

underground vault transformer: Transformer in vault beneath grade level to provide underground service to buildings.

V

very high radiation level: Nuclear power plant area that has a dose rate greater than 5 rem/hour at 30 cm from the radiation source.

vitrification: Method of changing radioactive material into a glass-like material to stabilize liquid radioactive wastes.

X

X ray: An electromagnetic radiation with a very short wavelength.

Y

yellowcake: Processed uranium ore ready for additional processing and enrichment.

Index

A

Accelerator, 53-54
Accident, nuclear power plant
  Chernobyl, 118-119
  Three Mile Island, 117-119
  Windscale, 117-118
AEC. *See* Atomic Energy Commission
Air monitor, 110-111, *111*
ALARA, 102, *103,* 143, 146
Alpha particle, 77, 77
Alpha radiation, 73-75, *75, 76,* 77-78,
  77
American National Standards Institute,
  98-99
ANSI. *See* American National
  Standards Institute
Antineutrino, 79
Area monitor, 111, *111*
As low as reasonably achievable. *See*
  ALARA
Atom
  defined, 4
  parts of, 4-7
    electron, 1, 4-6, *5, 6*
    neutron, 1, 4-6, *5, 6*
    nuclear forces, 6-7, 7
    nucleus, 4-5, *5, 6*
    proton, 1, 4-5, *5*
Atom bomb, 1, 15-17
  history of, 15-16
    Einstein, Albert, 15
    Fermi, Enrico, 16
    Hahn, Otto, 15
    Manhattan Project, 16-17, *17*
    Meitner, Lise, 15
    Roosevelt, Franklin D., 15
    Strassman, Fritz, 15
Atomic Energy Act of 1954, 93, 94-96,
  95, 143
  Atomic Energy Commission, 94, 96
  Code of Federal Regulations, 94-96,
  97
Atomic Energy Commission, 94, 96
Atomic number, 52
Atomic weight, 8, 52. *See also* Mass
  number

B

Becquerel, Henri, 14
Beta particle, 78-80, *78, 79*
  positrons, 78-79
  neutrino, 79
  antineutrino, 79
Beta radiation, 73-75, *75, 76,* 78-80,
  78, 79
Binding energy, 7, 53-54, *53, 54*
Blackout, 134
Boiling water reactor, 49, 67-68, *68,*
  122, *122*
Borosilicate glass, 168
Breeder reactor, 67-70, *71*
Brownout, 134
BWR. *See* Boiling water reactor

C

Canadian deuterium-uranium reactor,
  121-122, 124-125, *125*
CANDU reactor. *See* Canadian
  deuterium-uranium reactor
CFR. *See* Code of Federal Regulations
Chadwick, James, 13
Chain reaction, 16-17, *17,* 49, 55-57, *56*
  critical, 56
  subcritical, 56
  supercritical, 57, 65
  Manhattan Project, 16-17, *17*
Chemical compounds, 3
Chemical elements, 1, 3-4, *4,* 6-8
  defined, 3
  isotope, 7-8, *8*
Chemical mixtures, 3
Chernobyl, 118-119
Code of Federal Regulations, 94-96, *97*
Compounds, chemical, 3
Contaminated, 100-101, *101*
Control rods, 49, 59, *60,* 65, *65*
Control room, 121, 131, *132*
Coolant, 49, 59, *60,* 66
Cooling water provisions, 121,
  131-133, *133*
Core
  defined, 57-58
  reflector, 58-59